BookWare Companion Series™

Discrete-Time Control Problems using MATLAB and the Control System Toolbox

May we recommend these books in the BookWare Companion Series™

Djaferis	*Automatic Control: The Power of Feedback Using MATLAB®*
Frederick/Chow	*Feedback Control Problems Using MATLAB® and the Control System Toolbox*
Gardner	*Simulations of Machines Using MATLAB® and Simulink®*
Ingle/Proakis	*Digital Signal Processing Using MATLAB®*
Proakis/Salehi/Bauch	*Contemporary Communication Systems Using MATLAB®*
Schetzen/Ingle	*Discrete Systems Laboratory Using MATLAB®*
Strum/Kirk	*Contemporary Linear Systems Using MATLAB®*

DISCRETE-TIME CONTROL PROBLEMS

using

MATLAB®

and the

CONTROL SYSTEM TOOLBOX

Joe H. Chow
Rensselaer Polytechnic Institute

Dean K. Frederick
Saratoga Control Systems, Inc.

Nicholas W. Chbat
Mayo Clinic

Australia • Canada • Mexico • Singapore • Spain
United Kingdom • United States

Publisher: *Bill Stenquist*
Editorial Coordinator: *Valerie Boyajian*
Marketing Manager: *Tom Ziolkowski*
Project Manager, Editorial Production: *Mary Vezilich*
Print Buyer: *Vena Dyer*
Production Service: *Matrix Productions*
Compositor: *ATLIS Graphics & Design*
Printer: *Webcom, Ltd.*

Printed in Canada
1 2 3 4 5 6 7 06 05 04 03 02

For more information about our products, contact us at:
Thomson Learning Academic Resource Center
1-800-423-0563
For permission to use material from this text, contact us by: **Phone:** 1-800-730-2214
Fax: 1-800-730-2215
Web: http://www.thomsonrights.com

Library of Congress Control Number: 2002028634
ISBN 0-534-38477-3

Brooks/Cole–Thomson Learning
511 Forest Lodge Road
Pacific Grove, CA 93950
USA

Asia
Thomson Learning
5 Shenton Way #01-01
UIC Building
Singapore 068808

Australia
Nelson Thomson Learning
102 Dodds Street
South Melbourne, Victoria 3205
Australia

Canada
Nelson Thomson Learning
1120 Birchmount Road
Toronto, Ontario M1K 5G4
Canada

Europe/Middle East/Africa
Thomson Learning
High Holborn House
50/51 Bedford Row
London WC1R 4LR
United Kingdom

Latin America
Thomson Learning
Seneca, 53
Colonia Polanco
11560 Mexico D.F.
Mexico

Spain
Paraninfo Thomson Learning
Calle/Magallanes, 25
28015 Madrid, Spain

About the Series

Save time in the classroom and expand student learning with MATLAB®

Increasingly, the latest technologies and modern methods are crammed into courses already dense with important theory. As a result, many instructors now ask, "Are we simply teaching students the latest technology, or are we teaching them to reason?" We believe that these two alternatives need not be mutually exclusive. In fact, this series was founded on the belief that computer solutions and theory can be mutually reinforcing. Properly applied, computing can illuminate theory and help students to think, analyze, and reason in meaningful ways. It can also help them to understand the relationships and connections between new information and existing knowledge and to cultivate problem-solving skills, intuition, and critical thinking. The BookWare Companion Series was developed in response to this mission.

Specifically, the series is designed for educators who want to integrate computer-based learning tools into their courses, and for students who want to go further than their textbook alone allows. The former will find in the series the means by which to use powerful software tools to support their course activities without having to customize the applications themselves. The latter will find relevant problems and examples quickly and easily available and will have electronic access to them. Important for both educators and students is the premise on which the series is based: that students learn best when they are actively involved in their own learning. The BookWare Companion Series will engage them, provide a taste of real-life issues, demonstrate clear techniques for solving real problems, and challenge them to understand and apply these techniques on their own.

You can recommend ways to make the series even better, share your ideas about using technology in the classroom with your colleagues, suggest a specific problem or example for the next edition, or just let us know what's on your mind. We look forward to hearing from you, and we thank you for you continuing support.

Bill Stenquist	Publisher	bill.stenquist@brookscole.com
Tom Ziolkowski	Marketing Manager	tom.ziolkowski@brookscole.com

Contents

Preface

With the advances in microprocessor and digital signal processing (DSP) chips, digital control is becoming ubiquitous in everyday life. Control algorithms in advanced automotive, aircraft, and appliance applications can no longer be realistically and economically implemented using analog circuitry alone, and must be programmed as computer code stored in memory. The purpose of this book is to assist those who are studying the introductory aspects of digital control systems engineering by allowing them to use a digital computer to rapidly work a wide range of numerical problems so as to reinforce the learning process and gain deeper insight into control design. The book is built around illustrative examples that demonstrate the steps involved in the analysis and design process. The examples are followed by a variety of problems that span the spectrum from simple textbook-type reinforcement problems, to follow-up what-if problems, to realistic comprehensive problems and to open-ended exploratory problems. Two of the models used repeatedly in the comprehensive problems are the ball and beam system and the inverted pendulum system, which are popular in university control laboratories.

To accomplish this objective, the book uses the power of MATLAB and its Control System Toolbox. It is anticipated that this book and the accompanying MATLAB files that can be downloaded from the Brooks/Cole Web site will be used as a supplement to one of the many textbooks that cover the introductory aspects of digital control systems. It is not intended that this book be used as the sole reference for learning this material, as key results are summarized and illustrated but are not developed from basic principles.

A substantial amount of the material in this book is based on the notes and the MATLAB files developed for a studio course on discrete-time systems taught at Rensselaer Polytechnic Institute by the first author. In the studio mode, a class is held in a lecture/computer classroom, allowing lectures to be immediately followed by in-class MATLAB assignments for students to reinforce their learning of control concepts. For instructors teaching or considering teaching a digital-control course in a similar interactive learning setting, the M-files in this book can be readily adapted as in-class exercises.

The combination of this book, the accompanying files, and a suitable computer that runs MATLAB (Release 11 or 12) and the Control System Toolbox (Version 4 or higher) will provide the user with numerous opportunities for applying the techniques of linear discrete-time system analysis that form the basis for the analysis and design of feedback control systems. Because a powerful computer environment is almost always available, the user is no longer restricted to solving first-and second-order problems that can be worked out by hand. Nor is the user subjected to the drudgery of performing laborious calculations required to solve meaningful problems. Rather, the user can concentrate on interpreting the analysis and design results obtained with the computer, thereby enhancing the learning process.

The problems in this book are suitable for students taking a senior-level course on digital control systems. We assume the reader has already taken a first course on feedback control design of continuous-time systems. Many excellent textbooks can be used for a review of continuous-time feedback systems. A companion continuous-time control-system MATLAB problem book by two of the co-authors (DKF and JHC) of this text offers many helpful examples with available MATLAB code. Continuous-time systems are used in sampled-data control systems starting in Chapter 5. In addition to discrete-time system analysis methods based on transfer-function models, the book also includes problems on state-space models and design methods. As such, the book is also suitable as a supplement for students taking a graduate digital control system course. In addition, it can also be used as an overview and a guide for practicing engineers to gain familiarity with the computer-aided design of digital control systems. Because of the importance of digital signal processing concepts in practical control implementation, we discuss frequency response and digital filters in Chapter 6. This material fits quite logically into frequency-domain analysis and lead-lag compensator designs.

The material is organized in the same manner as many of the introductory digital control system textbooks. State-space modeling of systems is introduced early, in Chapter 4, to emphasize its importance in modeling real-world problems. However, for those following a syllabus without any state-space modeling content, Chapter 4 can be skipped without loss of continuity.

This book has been written based on the premise that learning control systems benefits from studying examples and their related MATLAB commands, followed by working on problems with increasing levels of complexity. Each example is designed to illustrate a specific concept and usually contains a script of the MATLAB commands used for the model creation and the computation. Some examples are followed by "what if" questions that allow the reader an immediate opportunity to answer related questions, to appreciate some of the more intricate parts of the concept. Then several reinforcement problems are provided to further apply the technique to textbook-type problems. At the end of each chapter, one or two exploratory problems are posed to practice the techniques on user-defined systems. From Chapter 2 on, several comprehensive problems are presented that involve the analysis and/or design of real-world systems whose models are presented in Appendix A. To solve them, the reader is required to apply the concepts discussed in the chapter. Each real-world model is analyzed and its control design considered with increasing sophistication as new techniques are introduced. For the ball and beam system and the inverted pendulum system, students with access to these experiments can directly apply their designs to verify the controller performance.

To make effective use of this book, the user is expected to have some familiarity with MATLAB, including data entry, plotting, and simple computations. The MATLAB M-files are available at the Brooks/Cole Web site http://www.brookscole.com/engineering_d/. The files can be downloaded and then used to solve all the examples, reinforcement problems, and comprehensive problems contained in the book. M-files that allow the user to do the exploratory problems are also given. The M-files also contain several special RPI functions that perform computations not found in either MATLAB or the Control System Toolbox.

ACKNOWLEDGMENTS

We would like to thank our editor, Bill Stenquist of Brooks/Cole, for his energetic support of this project for developing a digital control problem book as a companion to the continuous-time control problem book in the Brooks/Cole BookWare Companion Series. We are indebted to the following reviewers for providing inputs that were helpful in establishing the scope of this book:

Bonnie S. Heck, *Georgia Institute of Technology*
Jin Jiang, *University of Western Ontario*
Douglas K. Lindner, *Virginia Polytechnic Institute and University*
Hitay Ozbay, *The Ohio State University*
David G. Taylor, *Georgia Institute of Technology*
John P. Uyemura, *Georgia Institute of Technology*
Stephen Yurkovich, *The Ohio State University*

Special thanks are due to Professor Lester Gerhardt of Rensselaer, who taught many times with the first author the lecture version of the discrete-time systems course, and to the studio course students who provided feedback on improving the course. We are grateful for many former and current Rensselaer students who helped in the development of the MATLAB files. In particular, Shaopeng Wang and Jeff Breunstein developed many RPI functions, and Xuan Wei meticulously edited all the

M-files and cross-checked them with the examples. We would also like to thank Jacob Apkarian and Matthew Hodjera at Quanser Consulting, Inc., for discussions on the ball and beam and the inverted pendulum experiments used in the Comprehensive Problems.

At Brooks/Cole Publishing Company, we acknowledge the help of Mary Vezilich as Production Manager, with assistance from Sue Ewing, Permissions Editor, Tim Ziolkowski, Executive Marketing Manager, and Valerie Boyajian, Editorial Coordinator. Merrill Peterson from Matrix Productions coordinated the copy editing and production.

We are grateful to our employers, former and present (Rensselaer Polytechnic Institute, Unified Technologies, Saratoga Control Systems, General Electric Global Research, and Mayo Clinic), for the use of facilities and software that were essential to the preparation of this book.

Finally, we would like to thank our families for their support during the long hours we spent in preparing the text and the M-files.

J. H. Chow, Troy, New York
D. K. Frederick, Saratoga Springs, New York
N. W. Chbat, Rochester, Minnesota

To my father Er-Zhi and my mother Tian-Shen — JHC

To Phyllis....wife, mother, grandmother, first mate,
travel companion, and best friend — DKF

To my mother Marie, to the memory of my father Wadih,
and to Véra — NWC

1 *Introduction*

PREVIEW

This book contains examples and a variety of problems related to the analysis and design of discrete-time feedback control systems that can be solved with MATLAB® and the Control System Toolbox. As such, it serves as a companion to conventional discrete-time control systems textbooks, its examples and problems having been designed to reinforce the theory presented in these textbooks. Computer files that solve the examples and the problems can be found on the Brooks/Cole web site.[1] This first chapter describes the organization of the topics in the book and discusses how to use the book and its computer files.

[1] www.brookscole.com

MATLAB AND THE CONTROL SYSTEM TOOLBOX

The combination of MATLAB and the Control System Toolbox is the most widely used package for high-level, interactive control system design. *MATLAB*, which is derived from *MATrix LABoratory*, provides the underlying computational and graphical tools for handling real and complex scalars, vectors, and matrices. The *Control System Toolbox* is a set of functions written in the MATLAB language that makes it convenient to build the models and perform the analysis commonly done in control systems engineering. For example, with the Control System Toolbox commands, one can easily create system models as state-space or transfer-function objects and perform a variety of time-domain and frequency-response analyses, with graphical output when appropriate. These multifaceted model-building and analysis tools can then be used as the basis for a wide range of control system design methods. For the purpose of this book, we have implemented several additional functions for control system analysis, which we call RPI functions (for Rensselaer Polytechnic Institute). In Appendix D, we provide a list of MATLAB functions used in this text, together with a brief synopsis of each function and its origin. The example and problem MATLAB files are developed for version 6. Users of MATLAB version 5 should also be able to use the files, however.

We assume that the user has some experience with MATLAB, including its basic linear algebra features and plotting commands; therefore, we do not attempt to explain its features and uses. An excellent source for someone who has little or no experience with MATLAB is *Mastering MATLAB 6* by Hanselman and Littlefield (2001).[2]

If the reader knows MATLAB reasonably well, there should be no difficulty in using the Control System Toolbox commands as they are described and illustrated in the text, at least in their basic form. MATLAB and the Control System Toolbox have an excellent online help facility that allows the user to get a brief statement of the syntax of a function directly from the computer. The same online help facility is available for the RPI functions.

[2]See the bibliography at the end of the book for a detailed citation.

Because this book is intended to be a computer-aided design problems book for a first course in discrete-time control systems, we have used the models in the z-transform transfer-function form for most of the examples and problems. State-space models and design methods are contained in Chapters 4 and 10, respectively, however. We assume that a user of this book has taken a course in continuous-time control systems and is familiar with Laplace transforms. Therefore, we do not provide a review of continuous-time control system topics. Readers needing a review of these topics can look them up in standard textbooks or the companion to this book (Frederick and Chow, 2000).

The first part of the book, Chapters 2, 3, and 4, covers the building of models. The treatment includes both discrete-time transfer-function and discrete-time state-space models. The second part, Chapters 5, 6, and 7, introduces sampling, analysis techniques, and performance measures in the time domain and frequency domain. The third and last part, Chapters 8, 9, and 10, discusses discrete-time control design techniques. All these topics are covered in many standard control engineering textbooks for discrete-time systems. Table 1.1 cross-references the topics in this book with those found in a number of discrete-time control textbooks. The citations of the books in Table 1.1 are given in the bibliography.

	Chapter and Topic	*Franklin, Powell, & Workman*	*Kuo*	*Phillips & Nagle*	*Houpis & Lamont*	*Ogata*	*Jacquot*
2.	Single-block systems	4	2	2	4	2,5	2
3.	Multi-block systems	4	5	2	4	3	3
4.	State-space models	4	5	2	14	5	6
5.	Sampled-data systems	3,5,11	2,4,5	3,4,5	3,4,6	3,5	3
6.	Frequency response	6,7	8	7,11	5,6	3,4	5
7.	System performance	7	3,7,8	6,7	4,5	4	2,3
8.	Design in z-plane	3,7	7,10	8	5,11,13	4	4,5
9.	Frequency response design	7	10	8	13	4	5
10.	State-space design	8	6,10	9	15,17	6	8

TABLE 1.1 *Chapters in discrete-time control textbooks and their correspondence to the chapters of this book*

WAYS TO USE THIS BOOK

Each chapter in this book covers an important area of control-system analysis or design. Each section in a chapter deals with a key topic illustrated by one or more examples in which MATLAB is used. The important commands are often given in the solutions to the examples as "script" files. In MATLAB, these script files are called *M-files*, because they have the extension `.m`. In following the discussion of the example, the reader can type the MATLAB commands in the script files in an interactive manner or use M-files downloaded from the Brooks/Cole web site. A description of the M-files is provided below. To save space, most of the plotting commands are not listed in the script file in the example solution. They can be found in the example M-files, however.

Following each example, one or more what-if's may be posed, to lead the reader to examine the effect of variations in the key parameters. Each example is followed by several reinforcement problems that are similar in scope and solution approach. Comments that alert the reader to the special features of the problems are included, to enhance the learning experience. Partial answers to these problems are given at the end of the chapter.

After all the topics in a chapter have been covered, one or more exploratory problems may be posed, to allow the student to apply the MATLAB functions discussed in the chapter to models of any kind. Then several comprehensive problems based on models of physical systems are included. The models of the physical systems, together with descriptions of the systems, are given in Appendix A. Two of the models are experimental systems available in many university laboratories. These comprehensive problems tend to exercise the concepts and the MATLAB commands introduced in the chapter, to give the reader a more complete picture of the overall control analysis and design process.

The software obtainable from the web site will be of considerable interest to the reader because it contains the MATLAB M-files that run each of the examples and problems. These files are in MS-DOS® format and can be copied directly onto a PC. The MS-DOS files can also be read into a computer running either UNIX® or Linux®, using the proper file-transfer facilities. The M-files can be run on the corresponding computer platform having the full MATLAB version 5.0 or above and the Control System Toolbox version 4.1 or above. The M-files for the RPI functions are also available on the web site.

After studying a new topic in a conventional discrete-time control systems text, a reader should study the corresponding section in this book. Then the M-files containing the commands that solve the related examples and problems should be copied from the web site and run in MATLAB by typing its filename (without the `.m` extension). The convention for the example and problem M-files is as follows.

For all chapters, there is an M-file for each example. For instance, the M-file for Example 2.1 in Chapter 2 is named `ex2_1.m`,[3] and the M-file to generate Figure 2.1 is `makefig2_1.m`. The user can run these example M–files to generate all the MATLAB plots in this book. The M-files for the what-if questions start with the prefix `wif`, the reinforcement problems with `reinf`, and the comprehensive problems with `cp`. Therefore, the file `wif5_7.m` contains the MATLAB commands to solve the what-if question associated with Example 5.7, and the file `reinf2_4.m` solves Reinforcement Problem 2.4 in Chapter 2. The instruction for running the M-files for the exploratory problems is given with the problem statement.

To see the MATLAB commands used in an M-file, the `echo` command can be entered before executing the file. The reader may find it helpful to use the `diary` command to capture the contents of the MATLAB command window as the M-file executes. The resulting diary file can be printed and carefully studied afterwards.

In the M-files, we have included `pause` commands so that the display of computational results will not scroll off the computer display screen before the user has a chance to view them. When the program execution pauses, the characters `******>` are displayed. Simply press any key to continue. When the prompt `K>>` is displayed, the program is in the `keyboard` mode. This is implemented to allow the user to examine the newly created plot and, if desired, to make a hardcopy of the plot by using the `print` command. The user can also enter any MATLAB commands at this point. To continue the execution of the program, simply enter the `return` command.

For the user who wants to apply the tools discussed in this book to other systems, one convenient way is to use a copy of an appropriate M-file as a template. The system model in the template can be replaced with the new data, and additional MATLAB and Control System Toolbox commands can be inserted, if desired. Then the modified file can be run by typing its name without the `.m` extension. In this way, the user can quickly investigate the effects of changing the values of model parameters, controller gains, and inputs on whatever dynamic system is under consideration. When used properly, the combination of MATLAB and the many M-files included with this book can greatly accelerate the reader's understanding of control engineering.

[3] In the text, MATLAB commands and M-file names are printed in this typeface.

2 *Single-Block Models and Their Responses*

PREVIEW

The starting point of a control system design is the formulation of a system model, upon which Control System Toolbox functions can operate. Models of signals and fixed linear systems can be represented in several different mathematical forms. One of these forms is obtained by applying the z transform to the system, resulting in its *transfer function*, expressed as a ratio of polynomials in z. A second method for describing a system in MATLAB is in terms of the zeros, poles, and gain of its transfer function. In this chapter we show how both of these methods can be used to implement the model of a single subsystem, as would be represented by a single block in a block diagram. The analysis for a system with multiple blocks is considered in Chapter 3, and a third form of representing a fixed linear system, the state-space form, is discussed in Chapter 4.

TRANSFER FUNCTIONS

Consider a fixed single-input/single-output discrete-time linear system with input $u(k)$ and output $y(k)$ given by the difference equation

$$y(k+2) + 0.25y(k+1) - 0.375y(k) = 2u(k+1) + u(k) \tag{2.1}$$

where $k = 0, 1, 2, \ldots$ is the integer discrete-time variable. Applying the z transform to both sides of (2.1) with zero initial conditions, we obtain the transfer function of the system from the input $U(z)$ to the output $Y(z)$ in TF form as the ratio of polynomials.[1]

$$G(z) = \frac{2z+1}{z^2 + 0.25z - 0.375} \tag{2.2}$$

Alternatively, this transfer function can be expressed in terms of its zeros z_i, poles p_j, and gain K in the factored, or ZPK form, as

$$G(z) = \frac{2(z+0.5)}{(z+0.75)(z-0.5)} \tag{2.3}$$

which shows that $G(z)$ has a single zero at $z = -0.5$, two poles at $z = -0.75$ and 0.5, and a gain of 2.

If we know the numerator and denominator polynomials of $G(z)$, we can represent the model in MATLAB as a linear time-invariant (LTI) object in transfer function (TF) form by (i) creating a pair of *row* vectors containing the coefficients of the powers of z, in descending order, of the numerator and denominator polynomials, and (ii) using the `tf` command to create the TF object. For the transfer function in (2.2), we could enter `numG = [2 1]` and `denG = [1 0.25 -0.375]` to define the polynomials, followed by `G1 = tf(numG,denG,-1)` to create the LTI object. Assuming that we do not need to have the numerator and denominator polynomials available in the MATLAB workspace, we could build the TF object with the single statement `G1 = tf([2 1],[1 0.25 -0.375],-1)`.

The third argument in the `tf` command, `-1`, designates that the sampling period, T_s, is unknown. If T_s is known, then the `-1` is replaced with the value of the sampling period. If the third argument is omitted altogether, the assumption is that the system described by `numG` and `denG` is continuous and not discrete.

[1] $U(z)$ is the z-transform of the sequence $u(k)$, i.e., $U(z) = \mathcal{Z}\{u(k)\} = \sum_{k=0}^{\infty} u(k)z^{-k}$.

To illustrate, entering the command
`G1 = tf([2 1],[1 0.25 -0.375],-1)`
results in the following output to the screen:

```
Transfer function:
      2 z + 1
--------------------
z^2 + 0.25 z - 0.375

Sampling time: unspecified
```

For plotting purposes, a sampling period of unity will be assumed.

Alternatively, entering the command
`G2 = tf([2 1],[1 0.25 -0.375],0.25)`
produces the output

```
Transfer function:
      2 z + 1
--------------------
z^2 + 0.25 z - 0.375

Sampling time: 0.25
```

Note that the transfer function is the same in both cases, but the two LTI objects are not the same. The difference is due to the fact that the first system has its sampling time, T_s, unspecified, whereas the second system has $T_s = 0.25$ s.

Finally, if we enter the command `G3 = tf([2 1],[1 0.25 -0.375])`, we obtain the response

```
Transfer function:
      2 s + 1
--------------------
s^2 + 0.25 s - 0.375
```

which is expressed in terms of the complex variable s, rather than z. This difference, plus the absence of a sampling time in the display, indicates that we have defined the Laplace transform of a *continuous-time* system, not the z transform of a *discrete-time* system.

If the transfer function is known in terms of its zeros, poles, and gain, we can create the model as an LTI object in ZPK form by (i) entering *column* vectors for the zeros and poles, and entering the gain as a scalar and (ii) using the `zpk` command to create the ZPK object. To create the system described in (2.3) with an unspecified sampling period, we could enter the commands `zG = -0.5`, `pG = [-0.75; 0.5]`, `kG = 2`, and `G4 = zpk(zG,pG,kG,-1)`. Or we could use the single command `G4 = zpk(-0.5,[-0.75; 0.5],2,-1)`. In each case, the last argument, `-1`, specifies that the system is discrete time with an unspecified sampling period.

When a model has been described in MATLAB in either one of these forms, the Control System Toolbox provides the ability to convert from one form to the other. For example, if the system model has been created in TF form with its numerator and denominator polynomials as `Gtf`, we can create the ZPK form merely by entering `Gzpk = zpk(Gtf)`. Then, to determine the zeros, poles, gain, and sampling priod of $G(z)$, we can enter `[zz,pp,kk,Ts] = zpkdata(Gzpk,'v')`. In similar fashion, we could build the system in ZPK form as, say, `Sxx`, convert it to TF form via `Svv = tf(Sxx)`, and then enter `[nn,dd,Ts] = tfdata(Svv,'v')`. Finally, the last output argument, written as `Ts` for the sampling period, is optional.

There are several things to note about these commands. First, the name assigned to the TF or ZPK system is arbitrary, as evidenced by the use of `G1, G2, Gzpk, Gtf, Sxx`, and `Svv` as LTI object names. Second, the use of the string `'v'` (the single quotes around the letter `v` make it a string in MATLAB) as the second argument in the `tfdata` and `zpkdata` commands forces the outputs to be written as column vectors, provided that the system is single-input/single-output (SISO). For example, once the TF object `G1` has been created as shown earlier, the zeros, poles, and gain of $G(z)$ can be obtained by the command `[z,p,k] = zpkdata(G1,'v')` to extract the desired quantities into the MATLAB workspace. Without the string `'v'` in the `tfdata` or `zpkdata` commands, the output variables are returned as double-precision *cell arrays* with dimensions equal to the number of outputs times the number of inputs. For example, the poles from the `zpkdata` command are returned as a double-precision cell array, such as

```
pp =
        [2x1 double]
```

To display the numerical values of the two elements of `pp`, we enter the command `pp{:}`, which displays the numerical values of `pp` as

```
pp =
         0.9394
         0.7788
```

In general, the command `pp{i,j}` displays the poles of the system's transfer function from input `j` to output `i`.

Once we have the system model expressed as an LTI object—in either the TF or the ZPK form—we can obtain a graphical representation of the transfer function's poles and zeros by using the `pzmap` command from the Control System Toolbox. Assuming the name of the object is `G`, the command `pzmap(G)` displays an appropriate region of the z-plane with the zeros indicated by the symbol "o" and the poles by the symbol "×". Because the stability of a discrete-time system depends on the location of the transfer function's poles relative to the unit circle in the z plane, it is useful to show the unit circle on the pole-zero plot, which can be generated by the RPI function `ucircle`.

□ EXAMPLE 2.1 *Build $G(z)$ as a TF Object*

Create a TF object for the third-order system whose difference equation is

$$\begin{aligned} y(k+3) - 2.7y(k+2) + 2.42y(k+1) - 0.72y(k) \\ = \quad 0.1u(k+2) + 0.03u(k+1) - 0.07u(k) \end{aligned} \qquad (2.4)$$

with an unspecified sampling period. Then display the object's properties and extract the numerator and denominator polynomials from the TF object as row vectors. Also extract the zeros, poles, and gain.

Solution

Taking the z-transform of the difference equation with zero initial conditions, we obtain the following polynomial form of the transfer function

$$G(z) = \frac{0.1z^2 + 0.03z - 0.07}{z^3 - 2.7z^2 + 2.42z - 0.72}$$

The MATLAB commands in Script 2.1 will create $G(z)$ as a TF object, display all of its current properties (at the moment we are interested in only the first two), create the two row vectors, **nn** and **dd**, that contain the coefficients of the numerator and denominator polynomials, and draw the pole-zero plot shown in Figure 2.1. The system has poles at $z = 0.8$, 0.9, and 1.0, zeros at $z = 0.7$ and -1.0, and a gain of 0.1.

MATLAB Script

```
% Script 2.1:  Create G(z) as a TF object and determine some properties
numG  = [0.1 0.03 -0.07]          % enter transfer function numerator
denG  = [1 -2.7 2.42 -0.72]       % ...and denominator polynomials
%-- Create G(z) as discrete-time TF object with unspecified sampling period
G = tf(numG,denG,-1)
get(G)                            % show properties of the TF object
[nn,dd] = tfdata(G,'v')           % extract num & den from the TF object
[zz,pp,kk] = zpkdata(G,'v')       % extract data from ZPK object
ucircle,hold on                   % show unit circle in z-plane
pzmap(G),hold off                 % draw pole-zero plot
axis equal                        % equalize scales of real & imag axes
```

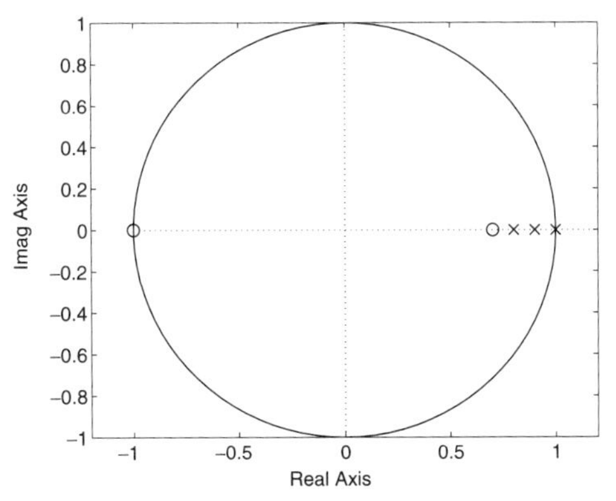

FIGURE 2.1 *Pole-zero plot for the system in Example 2.1* □

☐ EXAMPLE 2.2 *Convert from TF to ZPK Form*

Use the TF object created in Example 2.1 to obtain the model of the system described by (2.4) as a ZPK object. Then extract its zeros, poles, gain, and sampling time and show that they are the same as those of the TF object created in the preceding example. Leave the sampling time as unspecified.

Solution

We repeat the initial steps of Example 2.1 to create the TF object `G`. Then we convert it to a ZPK object named `GG` with the command `GG = zpk(G)`. To compare the zeros, poles, and gain of the two objects, we use the **`zpkdata`** command on both objects, but with different output arguments so that the results have distinct names. Doing this, we find that both systems have zeros at $z = -1.0$ and 0.7, poles at $z = 1.0, 0.9$, and 0.8, a gain of 0.1, and a sampling period of -1, which indicates that it is unspecified.

MATLAB Script

```
% Script 2.2:  Convert from TF to ZPK form
% Create G(z) as a TF object with Ts unspecified
G = tf([0.1 0.03 -0.07],[1 -2.7 2.42 -0.72],-1)
GG = zpk(G)                      % Convert G(z) into ZPK object
[zz,pp,kk,TTs] = zpkdata(GG,'v')% Extract poles, zeros, K, & Ts from ZPK form
[z,p,k,Ts] = zpkdata(G,'v')     % Extract poles, zeros, K, & Ts from TF form
```

☐

REINFORCEMENT PROBLEMS

For each of the systems whose transfer functions are given below as ratios of polynomials, implement the model in TF form, convert it to ZPK form, and draw a pole-zero plot with the unit circle included. The sampling periods are unspecified.

P2.1 Third-order system.

$$G(z) = \frac{1.25z^2 - 1.25z + 0.30}{z^3 - 1.05z^2 + 0.80z - 0.10}$$

P2.2 Fourth-order system.

$$G(z) = \frac{0.84z^3 - 0.062z^2 - 0.156z + 0.058}{z^4 - 1.03z^3 + 0.22z^2 + 0.094z + 0.05}$$

For each of the systems whose zeros, poles, and gain are given below, implement the model in ZPK form, convert it to TF form, and draw a pole-zero plot with the unit circle included. The sampling periods are unspecified.

P2.3 **Fourth-order system with complex poles.** $G(z)$ has zeros at $z = -0.2$ and 0.4, poles at $z = 0.6, -0.5 \pm j0.75$, and -0.3, and a gain of 150.

P2.4 **Fourth-order system with real poles.** $G(z)$ has zeros at $z = -0.3$ and $-0.4 \pm j0.2$, poles at $z = -0.6, 0.3, 0.5$, and 0.6, and a gain of 5.

RESIDUES AND UNIT-DELTA (IMPULSE) RESPONSE

Once a transform (or transfer function) $G(z)$ has been defined in MATLAB, operations can be performed to compute and display the corresponding discrete sequence $g(k)$, known as the *inverse z-transform.*

We can use the MATLAB `residue` function to compute the poles and their corresponding residues in a partial-fraction expansion, from which we can then obtain $g(k)$ in analytical form. When performing inverse z-transforms via the partial-fraction expansion, however, it is conventional to obtain terms of the form $Az/(z-p)$ rather than $A/(z-p)$, which is used when expanding Laplace transforms. The reason for this preference is that the former expression is the z-transform of the discrete-time function Ap^k, where p is a simple pole of the z-transform and $k = 0, 1, 2, \ldots$ is the discrete-time variable.

To be able to obtain the single z in each of the numerators of the expansion, we perform the partial-fraction expansion of $G(z)/z$ and then multiply through by z. The function $G(z)/z$ is just the transform $G(z)$ with an additional pole at $z = 0$, and we can obtain it from $G(z)$ in several ways. One of these is to multiply the denominator of $G(z)$ by z, which is easily accomplished in MATLAB by appending a `0` to the row vector containing the coefficients of the denominator polynomial.

Once $G(z)$ has been written as a partial-fraction expansion, the individual transform terms can be associated with their corresponding discrete-time functions, and the complete time function can be written as the sum of the individual terms, each of which can be associated with one of the poles of $G(z)$.

Alternatively, because $g(k)$ is the unit-delta response corresponding to $G(z)$, we can use the Control System Toolbox `impulse` function to obtain $g(k)$ in numerical form and plot it versus k. Both of these methods are illustrated in the following example.

☐ EXAMPLE 2.3 *Partial-Fraction Expansion and $g(k)$*

The transfer function of a time-invariant linear system is

$$G(z) = \frac{2z^2 - 2.2z + 0.56}{z^3 - 0.6728z^2 + 0.0463z + 0.4860} \tag{2.5}$$

and the sampling period is unity. Perform a partial-fraction expansion of $G(z)/z$, and use the results to write $G(z)$ as the sum of terms $Az/(z-p)$, each having a single pole and z in the numerator. Then build the system as a TF object and plot its impulse (unit-delta) response.

Solution

Because the `residue` command requires numerator and denominator polynomials as its two arguments, rather than an LTI object, we start by defining the row vectors `numG = [2 -2.2 0.56]` and `denG = [1 -0.6728 0.0463 0.4860]`, which contain the coefficients of the numerator and denominator polynomials, in descending order, left to right. As shown in Script 2.3, we use the `residue` command in the form `[rGoz,pGoz,other] = residue(numG,[denG 0])`, where the term `[denG 0]` has the effect of multiplying the denominator of $G(z)$ by z, as explained previously.

The result of this operation consists of three elements: (i) a column vector `rGoz` containing the residue at each of the poles of $G(z)/z$, (ii) a column vector `pGoz` containing the poles of $G(z)/z$, and (iii) a scalar `otherGoz` that is always empty, because the degree of the numerator of $G(z)/z$ will always be less than the degree of its denominator. Thus MATLAB sets it to `[ ]`, which is referred to as the *empty element.*

The `Goz` nomenclature is used to emphasize the fact that we are using $G(z)/z$ rather than $G(z)$ alone for the expansion. The poles and residues are summarized in Table 2.1. Note that the values of the complex poles and residues have been displayed in polar, rather than Cartesian, form. We have adopted this convention (it will be used for complex zeros, also) because, when dealing with discrete-time systems in the z plane, it is the *magnitudes* of the poles that provide the most information about the dynamic behavior of the system. When dealing with continuous-time systems in the s plane, it is more informative to display these quantities in terms of their real and imaginary parts.

Pole	*Residue*
0	1.1523
-0.6	-2.2410
$0.9\epsilon^{j0.7854}$	$0.5451\epsilon^{j0.0540}$
$0.9\epsilon^{-j0.7854}$	$0.5451\epsilon^{-j0.0540}$

TABLE 2.1 *Poles and zeros of $G(z)/z$ in Example 2.3*

From these MATLAB results, we can write the partial-fraction expansion of $G(z)/z$ as

$$\frac{G(z)}{z} = \frac{1.1523}{z} + \frac{-2.2410}{z+0.6} + \frac{0.5451\epsilon^{j0.0540}}{z-0.9\epsilon^{j0.7854}} + \frac{0.5451\epsilon^{-j0.0540}}{z-0.9\epsilon^{-j0.7854}}$$

which can be converted to the expansion of $G(z)$ by multiplying both sides by z. Doing this, we obtain this result

$$G(z) = 1.1523 + \frac{-2.2410z}{z+0.6} + \frac{(0.5451\epsilon^{j0.0540})z}{z-0.9\epsilon^{j0.7854}} + \frac{(0.5451\epsilon^{-j0.0540})z}{z-0.9\epsilon^{-j0.7854}} \tag{2.6}$$

where the constant term is due to a delta function that has the value 1.1523 for $k = 0$ and is zero for all other values of discrete time k. The unit-delta function is denoted by $\delta(k)$. The second term is the transform of a decaying exponential or power function, and the last two terms represent a decaying oscillation.

The unit-delta response can be computed by using the `impulse` command, which can take one of several forms. The simplest of these is to enter `impulse(G)`, where G is an LTI object. This command will cause a plot of $g(k)$ to be displayed, using a time interval selected by MATLAB.

Another option is to enter `impulse(G,dtime)`, where `dtime` is a *column* vector of discrete time sequence points that has been defined previously. If the sampling time is unspecified, `dtime` is interpreted as the number of samples. Otherwise, `dtime` should be of the form `[Ti:Ts:Tf]`, where `Ti` and `Tf` are the initial and final discrete-time values, and `Ts` is the sampling time.

A third option is to use a single left-hand argument, as in `y = impulse(G,dtime)` with the discrete-time sequence vector specified as before. This form of the command returns the output values for each discrete-time point in the column vector `y`, which can then be plotted versus discrete-time samples by entering `stem(dtime,y)` to show the points with lines back to the time axis, or `plot(dtime,y,'o')` for a plot that uses a symbol, in this case an open circle, without the stem lines. What we should *not* do when plotting the response of a discrete-time system is use the `plot` command without one of the special symbols. This is because the result would be a plot having a continuous curve, without the discrete-time values shown as individual points. In such cases, the viewer of the plot is likely to form the mistaken impression that the results are those of a continuous-time system, which is not the case.

A fourth option is to use two left-hand arguments, as in `[y,dt] = impulse(G)`, where `y` is the response just described and `dt` is the corresponding discrete-time sequence. The unit-delta response for $G(z)$ appears in Figure 2.2.

MATLAB Script

```
% Script 2.3:  Partial-fraction expansion and unit delta response
numG = [2 -2.2 0.56]                         % define numerator of G(z)
denG = [1 -0.6728 0.0463 0.4860]             % define denominator of G(z)
[rGoz,pGoz,other] = residue(numG,[denG 0])   % residues & poles of G(z)/z
[mag_pGoz,theta_pGoz] = xy2p(pGoz)           % magnitudes & arguments of poles
[mag_rGoz,theta_rGoz] = xy2p(rGoz)           % magnitudes & arguments of residues
G = tf(numG,denG,1)                          % G(z) as TF object with Ts = 1 s
[y,k] = impulse(G);                          % compute unit delta response
stem(k,y,'filled')                           % plot response
```

WHAT IF? Try changing the coefficients of the numerator of $G(z)$ while leaving the denominator alone. Look at the effects on the poles, zeros, gain, residues, and unit-delta response, and verify that all of these parameters are affected except the poles. In other words, by changing the numerator of $G(z)$ but not its denominator, you affect the weighting of the system's mode functions but not the mode functions themselves. You can use the file `wif2_3.m` to get started. ■

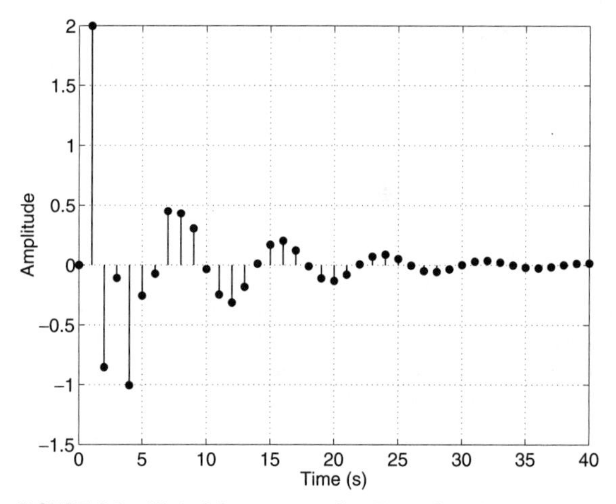

FIGURE 2.2 *Unit delta response for Example 2.3*

□

For each of the difference equations that follow, write the transfer function $G(z)$ as a ratio of polynomials. Then use MATLAB to find the zeros, poles, and gain of $G(z)$, to perform a partial-fraction expansion, and to compute and plot the impulse response. Assume the sampling period is unity.

P2.5 Third-order system.

$$\begin{aligned} y(k+3) + 0.7y(k+2) + 0.5y(k+1) + 0.1y(k) \\ = \quad 0.2u(k+2) + 1.1u(k+1) + 0.5u(k) \end{aligned}$$

P2.6 Fourth-order system.

$$\begin{aligned} y(k+4) - 1.3y(k+3) + 0.455y(k+2) + 0.0628y(k+1) - 0.037y(k) \\ = \quad 0.3u(k+2) + 0.4u(k+1) + 0.5u(k) \end{aligned}$$

TIME RESPONSE DUE TO DISTINCT POLES

In addition to computing the time response of the output, we can use MATLAB to evaluate and display the time response due to an individual real pole or to a pair of complex poles. Recall that a real pole at $z = p$ with residue r results in the time function $y(k) = rp^k$ for $k \geq 0$. If a pole is complex, say at $z_1 = a + jb$, its complex conjugate $z_2 = a - jb$ is also a pole. As described in the previous example, when dealing with discrete-time systems, it is more informative to represent the complex pairs of poles in polar form, in which case we write $z_1 = M\epsilon^{j\omega}$ and $z_2 = M\epsilon^{-j\omega}$. The residue at z_1 is a complex number, say $r_1 = A\epsilon^{j\phi}$, and the residue at z_2 is the complex conjugate of r_1, namely $r_2 = A\epsilon^{-j\phi}$. It can be shown that for $k \geq 0$, the time response due to this pair of complex poles is $y(k) = 2AM^k \cos(\omega k + \phi)$.

To simplify the calculation of these responses we have created two RPI functions, `rpole2k` and `cpole2k`. The former function has as its input arguments the value of the real pole, its residue, and a time sequence vector. The function returns a column vector of time response values, one per time sample. The latter function has the value of the complex pole with the *positive* imaginary part as its first input argument, the residue at this pole as the second argument, and a time sequence vector as the final argument.

□ EXAMPLE 2.4 *Responses Due to Individual Poles*

Use the poles and residues of the transfer function $G(z)/z$ for the $G(z)$ given in Example 2.3 to display the components of $g(k)$ due to the real poles at $z = 0$ and $z = -0.6$ and the complex poles at $z = 0.90\epsilon^{\pm j0.7854}$. Verify that the sum of all these responses is the same as the unit-delta response shown in Figure 2.2.

Solution

Referring to Table 2.1, we use the function `cpole2k` with the pole $z = 0$ and the residue $r = 1.1523$ and then with the pole $z = -0.6$ and the residue $r = -2.2410$ to obtain the responses due to the two real poles of $G(z)/z$. Each resulting time sequence is the inverse z-transform of the first and second terms in (2.6). For the last two terms in (2.6), which result from the complex poles, we use the pole having the positive imaginary part, namely $z = 0.90\epsilon^{j0.7854}$, and its residue $r = 0.5451\epsilon^{j0.0540}$. The MATLAB instructions that accomplish these steps and plot the responses are shown in Script 2.4, and the resulting responses are shown in Figure 2.3. The response due to the pole at $z = 0$ is nonzero only at $k = 0$, and the response due to the negative real-axis pole at $z = -0.6$ is oscillatory as it alternates in sign.

MATLAB Script

```
% Script 2.4:  Responses due to individual poles
numG = [2 -2.2 0.56]                      % create G(z)
denG = [1 -0.6728 0.0463 0.4860]
[rGoz,pGoz,other] = residue(numG,[denG 0]) % residues, poles of G(z)/z
[mag_pGoz,theta_pGoz] = xy2p(pGoz)        % show poles in polar form
[mag_rGoz,theta_rGoz] = xy2p(rGoz)        % show residues in polar form
dtime = [0:30];                           % sampled time data sequence
ycmplx = cpole2k(pGoz(1),rGoz(1),dtime);  % response due to complex poles
yreal1 = rpole2k(pGoz(3),rGoz(3),dtime);  % response due to 1st real pole
yreal2 = rpole2k(pGoz(4),rGoz(4),dtime);  % response due to 2nd real pole
ytot = ycmplx + yreal1 + yreal2;          % ytot(k) is sum of the three
%----- plot individual responses as separate plots
subplot(3,1,1)          %----- Response due to complex poles -----
stem(dtime,ycmplx,'m','filled'),grid
subplot(3,1,2)          %----- Response due to real poles ------
stem(dtime,yreal1,'g','filled'),grid
hold on
stem(dtime,yreal2,'r')
hold off
subplot(3,1,3)          %----- Complete response -------
stem(dtime,ytot,'b','filled'),grid
```

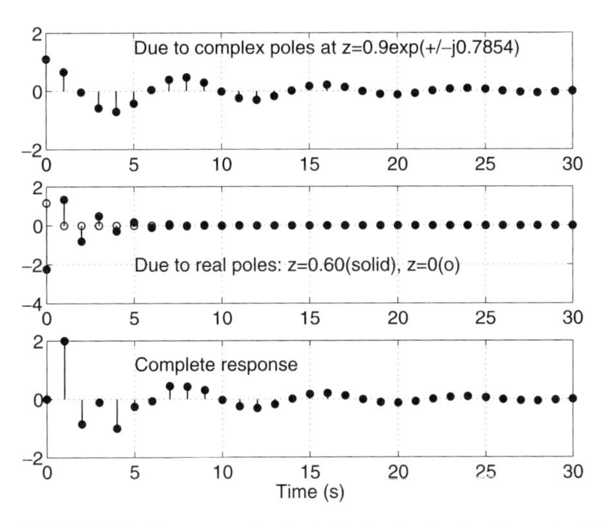

FIGURE 2.3 *Responses due to individual poles for Example 2.4*

C R O S S - C H E C K Verify that the complete response plot in Figure 2.3, which is the sum of the individual responses, agrees with the response shown in Figure 2.2 that was obtained by using the `impulse` command. ■

□

REINFORCEMENT PROBLEMS

P2.7 Third-order system with responses due to individual poles. For the transfer function $G(z)$ used in Problem 2.5, compute and plot the individual responses due to the real and complex poles. Verify that the sum of these responses agrees with the unit-delta response $g(k)$ from Problem 2.5.

P2.8 Fourth-order system with responses due to individual poles. Repeat the steps of Problem 2.7 as applied to the transfer function $G(z)$ in Problem 2.6.

☐ EXAMPLE 2.5 *Response Due to Complex Poles on the Unit Circle*

Form a transfer function, $G(z)$, with the complex poles $z = \epsilon^{\pm j\pi/3}$ on the unit circle, a zero at $z = -0.5$, and a gain of unity. Plot the first 20 seconds of the impulse response when the sampling period is unity. Then redo the plot when the sampling period is 0.5 s. Explain how the periods of the responses are related to the values of the zero and the poles.

Solution

As shown in Script 2.5, we write the poles of $G(z)$ as $z = A\epsilon^{\pm j\pi/3}$, where $A = 1$, and use the `conv` command to multiply the two first-order polynomials in z to get the denominator of $G(z)$. Then we build the system as a TF object with $T_s = 1$ s and use the `impulse` command to obtain the plot shown in Figure 2.4(a). Note the undamped oscillations due to the complex poles on the unit circle. Finally, we change the sampling period to 0.5 s, rebuild the system model, recompute the impulse response, and plot the result, obtaining Figure 2.4(b). Comparing the two figures, we see that by keeping the same transfer function but halving the sampling period, we get exactly the same sequence of values, but at twice the rate.

MATLAB Script

```
% Script 2.5:  Response due to poles on the unit circle
Ts = 1
numG = [1 0.5]                                      % numerator of G(z)
denG = conv([1 -exp(i*pi/3)],[1 -exp(-i*pi/3)])  % denominator of G(z)
G1 = tf(numG,denG,Ts)                  % build G(z) as TF object for Ts = 1 s
[y1,k1] = impulse(G1,20);              % compute response and time vector
stem(k1,y1,'filled'),grid              % plot response for Ts = 1 s
%------ change sampling period to Ts = 0.5 s --------
Ts = 0.5
G05 = tf(numG,denG,Ts)                 % rebuild G(z) as TF object for Ts = 0.5 s
[y05,k05] = impulse(G05,20);           % compute response and time vector
stem(k05,y05,'filled'),grid            % plot response for Ts = 0.5 s
```

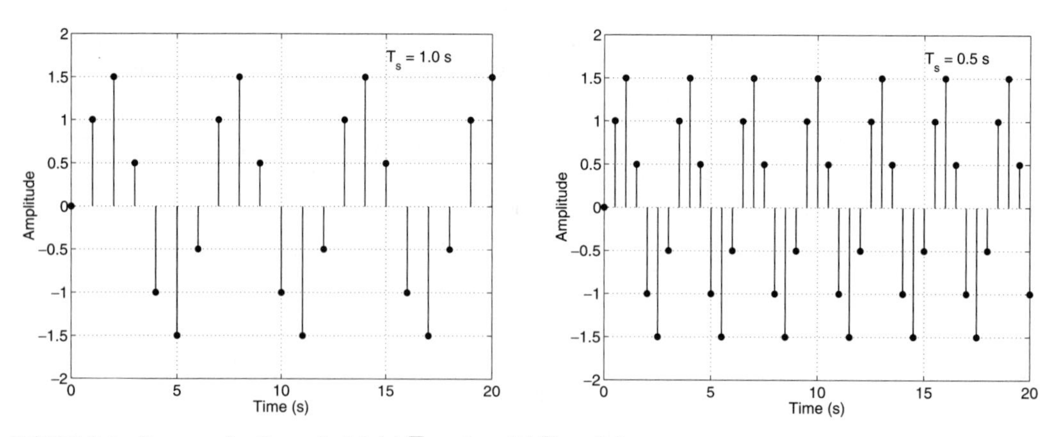

FIGURE 2.4 *Response for Example 2.5 (a) $T_s = 1$ s; (b) $T_s = 0.5$ s*

WHAT IF? Change the coefficient A in the previous example to values slightly less than 1 and then to values slightly greater than 1. Plot the corresponding impulse response in each case. Use a sampling period of $T_s = 1$ s. ■

□

TIME RESPONSE DUE TO REPEATED POLES

Up to this point, the discussion has been restricted to distinct poles, either real or complex. For a repeated pole, there is more than one term in the time response, with the number of terms depending on the multiplicity of the pole. For a pole at $z = p$ of multiplicity $m = 2$, we can write the z-transform of the response as

$$G(z) = \frac{Az}{(z-p)^2} + \frac{Bz}{z-p} + \text{ terms involving other poles} \tag{2.7}$$

where

$$A = \left[(z-p)^2 \frac{G(z)}{z} \right] \Bigg|_{z=p} \quad \text{and} \quad B = \left\{ \frac{d}{dz} \left[(z-p)^2 \frac{G(z)}{z} \right] \right\} \Bigg|_{z=p}$$

For $k \geq 0$, the corresponding unit delta response is

$$g(k) = Akp^{k-1} + Bp^k + \text{ terms involving other poles} \tag{2.8}$$

Similar relationships exist for poles of multiplicity three or higher, but they are seldom needed and will not be covered here. Nor will repeated complex poles be covered. The `residue` command is able to compute the coefficients for repeated poles, and for $m = 2$ they are listed as B followed by A.

□ EXAMPLE 2.6 *Repeated Poles*

For the system whose difference equation is

$$y(k+2) - 1.8y(k+1) + 0.81y(k) = 3u(k+1) - 1.2u(k)$$

with a sampling period $T_s = 1$ s, do a partial-fraction expansion of $G(z)/z$ and use the result to write the unit-delta response as the sum of three individual functions of the discrete-time variable k. Plot the unit-delta response.

Solution

The transfer function $G(z)/z$ is

$$\frac{G(z)}{z} = \frac{3(z-0.4)}{z^3 - 1.8z^2 + 0.81z} = \frac{3(z-0.4)}{z(z-0.9)^2} = \frac{A}{(z-0.9)^2} + \frac{B}{z-0.9} + \frac{C}{z}$$

which has a pole at $z = 0.9$ of multiplicity two and a simple pole at $z = 0$. Use of the `residue` command results in the column vectors

```
rGoz = 1.4815      and      pGoz = 0.9000
       1.6667                      0.9000
      -1.4815                           0
```

and the scalar `other = [ ]`. Associating the first element of `rGoz` with B, the second element with A, and the third with C, we can write

$$\frac{G(z)}{z} = \frac{1.6667}{(z-0.9)^2} + \frac{1.4815}{z-0.9} - \frac{1.4815}{z}$$

From (2.7) and (2.8), where there are no other poles, we see that for $k \geq 0$, the unit delta response is

$$g(k) = 1.6667k(0.9)^{k-1} + 1.4815(0.9)^k - 1.4815\delta(k)$$

MATLAB Script

```
% Script 2.6:  Repeated poles
numG = [3 -1.2]
denG = [1 -1.80 0.81]
[rGoz,pGoz,other] = residue(numG,[denG 0]) % calculate residues & poles
```

WHAT IF? Generate and plot the functions 0.9^k and $k0.9^{k-1}$ for $k = 0, 1, 2, \ldots, 50$. ■

□

REINFORCEMENT PROBLEMS

Use MATLAB to obtain a partial-fraction expansion of the system's transfer function and to write the impulse response as a sum of individual time functions. Then plot the impulse response. Use a sampling period of unity.

P2.9 Third-order system with one repeated pole.

$$y(k+3) + 0.4y(k+2) - 0.35y(k+1) - 0.15y(k) = 0.6u(k+1) + 0.4u(k)$$

P2.10 Fourth-order system with two repeated poles.

$$\begin{aligned} y(k+4) + 0.6y(k+3) - 0.11y(k+2) - 0.06y(k+1) + 0.01y(k) \\ = 0.3u(k+2) + u(k+1) - 0.5u(k) \end{aligned}$$

The z-transform of a system's unit-step response is the product of the system's transfer function, $G(z)$, and $z/(z-1)$, the transform of the unit-step function. The poles of the resulting transform are the poles of $G(z)$ and a pole at $z = 1$ (due to the unit-step input). The zeros of the step response are the zeros of $G(z)$ and a zero at $z = 0$. The step response can be computed and plotted using the `step` command from the Control System Toolbox, which has the same options for plotting and returning numerical values as the `impulse` command does.

The system gain at $z = 1$ is $G(1)$ and is known as the DC gain. We can compute the DC gain directly from either the TF or ZPK object using the Control System Toolbox `dcgain(G)` command.

In the following example, we use MATLAB to construct the z-transform of a step response, plot the response with the `impulse` command, and compare the result with a plot obtained using the `step` command. We also illustrate the use of the initial- and final-value theorems.

□ EXAMPLE 2.7 *Step Response from $G(z)$*

For the transfer function $G(z)$ used in Example 2.3 and given by (2.5), obtain a plot of the step response by adding a pole at $z = 1$ and a zero at $z = 0$ to $G(z)$ and using the `impulse` command to plot the inverse z-transform. The sampling period is $T_s = 1$ s. Compare the response with that obtained using the `step` command applied to $G(z)$. Determine the system's DC gain and verify that the result of applying the final-value theorem agrees with your results.

Solution

Because the system's transfer function is the same as that in the previous example, we use the same values for `numG` and `denG`. To add a pole at $z = 1$, we multiply the denominator polynomial by $z - 1$, and to add a zero at $z = 0$, we multiply the numerator by z. One method of accomplishing this in MATLAB is to use the `conv` command, which multiplies two polynomials by performing a discrete convolution of the row vectors containing their coefficients. To multiply the denominator of $G(z)$ by the polynomial $z - 1$, which in MATLAB is represented by the row vector `[1 -1]`, we enter `denGstep = conv(denG,[1 -1])`. In a similar way, to multiply the numerator polynomial by z, we enter `numGstep = conv(numG,[1 0])`. Note that the result of this last MATLAB statement can also be obtained as `numGstep = [numG 0]`.

Having obtained the transform of the unit-step response in terms of the row vectors `numGstep` and `denGstep`, we create the TF object `Gstep` and use the `impulse` command to get the inverse transform, which is the step response shown in Figure 2.5. Alternatively, we could have used the TF object `G` constructed from the polynomials `numG` and `denG` and the `step` command to generate the same plot. The calculation of the DC gain gives a value of 0.4188, which agrees with $G(1)$ as obtained by letting $z = 1$ in (2.5). These commands are contained in Script 2.7.

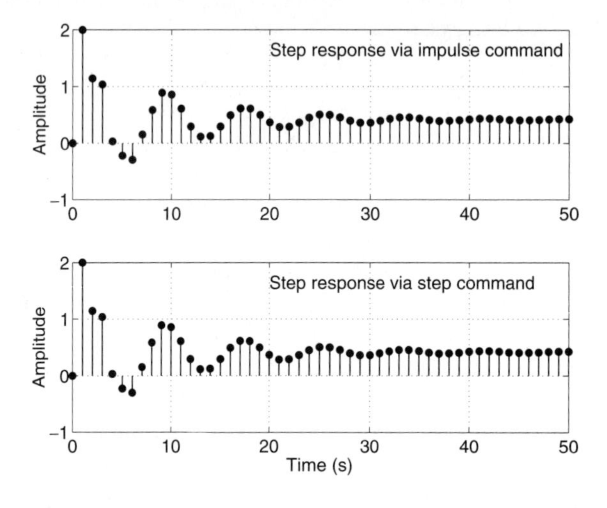

FIGURE 2.5 *Step response for Example 2.7*

MATLAB Script

```
% Script 2.7:  Obtain step response from G(z)
numG = [2 -2.2 0.56]                     % create G(z)
denG = [1 -0.6728 0.0463 0.4860]         % ...as TF object
G = tf(numG,denG,1)                      % ...with unit sampling period
numGstep = [numG 0]                      % add a zero at z = 0 to G(z)
denGstep = conv(denG,[1 -1])             % add a pole at z = 1 to G(z)
Gstep = tf(numGstep,denGstep,1)          % create zG(z)/(z-1) as TF object
dtime = [0:50];                          % define discrete time samples
yi = impulse(Gstep,dtime);               % step response via impulse cmd
subplot(2,1,1)
stem(dtime,yi,'filled')                  % plot impulse response of Gstep(z)
ys = step(G,dtime);                      % step response via step command
subplot(2,1,2)
stem(dtime,ys,'filled')                  % plot step response of G(z)
dcgain(G)                                % DC gain of G(z)
```

If $y_s(k)$ denotes the system's unit-step response, we can see from Figure 2.5 that $y_s(0) = 0$, and we can use the command `ys_ss = ys(max(dtime))` to determine that the steady-state response is $y_s(\infty) = 0.4188$. Another way to obtain this value is to use `ys_ss = ys(end)`, which returns the last element of the output vector `ys`. To verify these values analytically, we write the z-transform of the step response as

$$Y_s(z) = \frac{z}{z-1}G(z) = \frac{z}{z-1} \cdot \frac{2z^2 - 2.2z + 0.56}{z^3 - 0.6728z^2 + 0.0463z + 0.486}$$

The initial-value theorem gives

$$y_s(0) = \lim_{z \to \infty} Y_s(z) = 0$$

because the denominator has a higher degree in z than does the numerator. The final-value theorem gives

$$\begin{aligned} y_s(\infty) &= \lim_{z \to 1} \frac{z-1}{z} Y_s(z) = \lim_{z \to 1} \frac{z-1}{z} \cdot \frac{z}{z-1} G(z) \\ &= \lim_{z \to 1} G(z) = G(1) = 0.4188 \end{aligned}$$

WHAT IF? Find and plot the response of the system in Example 2.3 to the unit-ramp function by adding two poles at $z = 1$ and a zero at $z = 0$ to $G(z)$. ■

□

REINFORCEMENT PROBLEMS

P2.11 Step response. Obtain a plot of the unit-step response for the system described in Problem 2.5. Use the final-value theorem to verify the steady-state value of the response.

P2.12 Another step response. Repeat the steps of Problem 2.11 for the system described in Problem 2.6.

RESPONSE TO A GENERAL INPUT

In addition to computing and plotting the impulse and step responses of a system, MATLAB can be used to simulate and display the response to general functions of time. This is done with the `lsim` command in a variety of ways. In its simplest form, the user specifies the system as an LTI object (in either TF or ZPK form), a vector of input values, and a vector of discrete-time points.

If the `lsim` command, which performs linear simulation, is given with no output variable, the plot of the response is drawn but no numerical values are returned. A more useful way to use this command is to specify the system output (`y`) as a result and then to plot both the output and the input (`u`) versus discrete time. We illustrate the use of `lsim` in the following example by solving for the zero-state response to an input signal that is piecewise constant.

□ EXAMPLE 2.8 *Response to Piecewise-Constant Input*

Find the zero-state response of the system discussed in Examples 2.3, 2.4, and 2.7 to the input

$$u(k) = \begin{cases} 0 & k < 0 \\ 2 & 0 \leq k < 10 \\ 0.5 & k \geq 10 \end{cases}$$

over the interval $0 \leq k \leq 40$. The sampling period is $T_s = 0.2$ s.

Solution

Once the system has been defined as an LTI object, we need to create vectors of time points and input values. The only constraints are that the discrete-time points must have a uniform spacing equal to the system's sampling period and the two vectors must have the same number of elements. We start by defining `dtime` to be a *column* vector that has time increments of T_s and a large enough maximum time to show all of the transients. The command `dtime = [0:Ts:8]'` accomplishes this for the system and input signal under consideration.

The input vector `u` can be created in a number of ways. As shown in Script 2.8, we first use the `ones` command to make `u` a vector having the same dimension as `dtime`, but with all of its values equal to 2.0. Then we use the `find` command to determine which elements of `u` correspond to $k \geq 10$ and set these values to 0.5. The values of the response are computed by the `lsim` command and are stored in the column vector `y`. Then both `y` and `u` are plotted versus `dtime`, where the commands `hold on` and `hold off` are used to allow the overplotting. The result is shown in Figure 2.6.

MATLAB Script

```
% Script 2.8:  Response to piecewise-constant input
numG = [2 -2.2 0.56]                        % numerator of G(z)
denG = [1 -0.6728 0.0463 0.4860]            % denominator of G(z)
Ts = 0.2                                    % sampling period
G = tf(numG,denG,Ts)                        % create TF object G(z)
dtime = [0:Ts:8]';                          % discrete-time samples
u = 2.0*ones(size(dtime));                  % start with all input values = 2.0
ii = find(dtime>=2.0)                       % use find cmd to get indices of
u(ii) = 0.5;                                % ...elements to be changed
y = lsim(G,u,dtime);                        % compute response due to input u
stem(dtime,y,'filled'),grid                 % plot response of G(z) due to u
hold on
plot(dtime,u,'o')                           % plot input
hold off
text(2.3,-1.8,'output')                     % add labels to plot
text(1.6,2.3,'input')
```

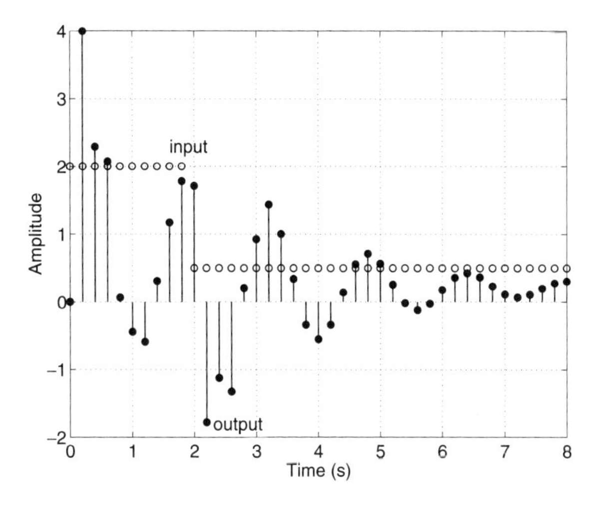

FIGURE 2.6 *Input and response of system in Example 2.8*

W H A T I F ? Suppose the input to the system is the one used in the example, but it is delayed by 1 s. Modify the commands in Script 2.8 to compute and plot the response, along with the new input. ■

□

REINFORCEMENT PROBLEMS

P2.13 Response to ramp-to-constant input. Compute and plot the zero-state response of the system described in Problem 2.5 when the input is the discrete-time function $u_1(k)$ obtained by sampling the function $u_1(t)$ shown in Figure 2.7 with the sampling period $T_s = 0.2$ s.

P2.14 Response to one cycle of a sampled sine wave. Compute and plot the zero-state response of the system described in Problem 2.6 when the input is the discrete-time function $u_2(k)$ obtained by sampling the function $u_2(t)$ shown in Figure 2.7 with the sampling period $T_s = 0.2$ s.

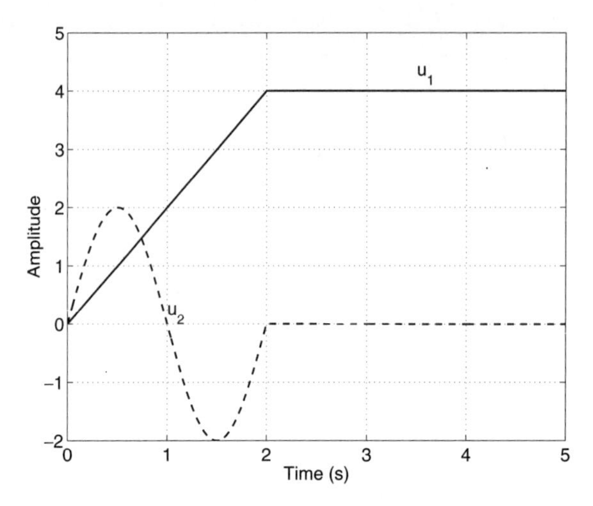

FIGURE 2.7 *Continuous input signals for Problems 2.13 and 2.14*

POLES AND STABILITY

If p_i is a pole of $G(z)$, then the natural, or zero-input, response of $G(z)$ consists of the *mode functions* ${p_i}^k$ if p_i is distinct and $k^q {p_i}^k$, $q = 0, 1 \ldots,$ $r-1$ if p_i has multiplicity $r > 1$. Thus the natural response decays to zero if $|p_i| < 1$ for $i = 1, \ldots, n$, that is, if all the poles are inside the unit circle in the z-plane. Such a system is said to be *asymptotically stable.* If all the poles are inside the unit circle except for distinct poles that lie on the unit circle, the natural response contains undamped sinusoids or a nonzero constant, and the system is said to be *marginally stable.* If some of the poles lie outside the unit circle in the z-plane or on the unit circle with multiplicity greater than one, the natural response is unbounded and the system is said to be *unstable.*

The following example illustrates two ways to determine a system's stability: (i) by using the `roots` command on the denominator of $G(z)$ or (ii) by using the `pole` command on the LTI object `G`.

☐ EXAMPLE 2.9 *Poles and System Stability*

Find the poles of the transfer function

$$G(z) = \frac{3z^2 + 1.8z + 1.08}{z^4 - 1.25z^3 + 0.495z^2 - 0.0035z - 0.1862}$$

and determine whether or not the system is stable. Also, determine the stability by using the `roots` command on the denominator polynomial. Plot the poles and zeros of $G(z)$ in the z-plane. Finally, demonstrate the system stability by simulating its response to the unit-delta function. The sampling period is unity.

Solution

We build the system model as a TF object by defining the row vectors `numG = [3 1.8 1.08]` and `denG = [1 -1.25 0.495 -0.0035 -0.1862]` to represent the numerator and denominator of $G(z)$, respectively. Then, we can build the TF object `G` and use it as the input argument of the `pole` command. Using the `pole(G)` command, we find the poles of $G(z)$ to be $z = 0.95$, $z = 0.35 \pm j0.6062$ (which can also be written as $0.7\epsilon^{\pm j1.0472} = 0.7\epsilon^{\pm j\pi/3}$), and $z = -0.4$. To evaluate the poles of $G(z)$, we can also use the `roots` command on the denominator of its transfer function by entering `ppG = roots(denG)`. The result is the same as that obtained using the `pole` command. Since all the poles have magnitudes of less than one, that is, they are all inside the unit circle, the system is asymptotically stable. The unit delta response can be found from the `impulse(G)` command and will decay to zero, as shown in Figure 2.8. It is possible to draw stem plots with dotted, rather than solid, stems. To do this for the variable `y`, we insert `':'` in the third argument, as in `stem(k,y,':o')`. The details are given in Script 2.9.

MATLAB Script

```
% Script 2.9:  find poles and impulse response
numG = [3 1.8 1.08]                    % generate G(z) as TF object
denG = [1 -1.25 0.495 -0.0035 -0.1862]
G = tf(numG,denG,1)                    % sampling period = 1.0 s
pG = pole(G)                           % poles of G via pole command
ppG = roots(denG)                      % poles of G via roots command
ucircle,hold on                        % show unit circle in z-plane
pzmap(G),hold off                      % draw pole-zero plot
axis equal                             % equalize scales of real & imag axes
[y,k] = impulse(G);                    % compute delta response of G(z)
stem(k,y,':o')                         % plot with dotted stems & o's
```

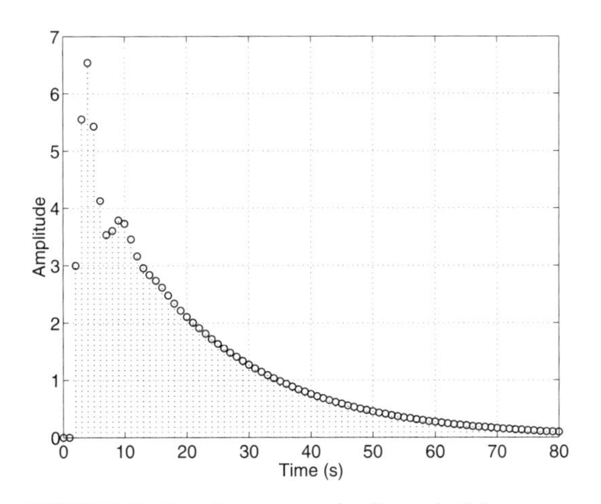

FIGURE 2.8 *Impulse response for Example 2.9*

CROSS-CHECK Verify that the poles of $G(z)$ are not affected if the sampling period T_s is changed. Also, show that the poles of $G(z)$ can be found by using the `damp(G)` command, which computes the natural frequencies and damping factors of LTI system poles. ■

WHAT IF?

a. Repeat Example 2.9 with the denominator of $G(z)$ changed to $z^4 - 1.3z^3 + 0.51z^2 - 0.014z - 0.196$. This system now has a pole on the unit circle, in place of the pole at $z = 0.95$ in the example.

b. Repeat Example 2.9 with the denominator of $G(z)$ changed to $z^4 - 1.5z^3 + 0.57z^2 - 0.056z - 0.2352$. This system now has a pole outside the unit circle in the z-plane. ■

□

REINFORCEMENT PROBLEMS

In each of the following problems, find the poles of the given transfer function $G(z)$, and determine the stability of the system. Verify the stability by simulating the impulse response. If a system is unstable, provide a discrete-time vector from $k = 0$ to 20 samples as the second input to the `impulse` function so the scaling of the plot allows the initial transients to be distinguished. Use a sampling period of unity in all cases.

P2.15 Fourth-order system.

$$G(z) = \frac{0.1z^2 + 0.3z + 0.4}{z^4 + 0.8z^3 + 0.3z^2 + 0.76z + 0.8}$$

P2.16 Pole at $z = -1$.

$$G(z) = \frac{z^3 + 0.2z^2 + 0.7z + 0.2}{z^4 + 0.1z^3 - 0.62z^2 + 0.24z - 0.04}$$

P2.17 Complex poles on the unit circle.

$$G(z) = \frac{z + 0.3}{z^3 - 2.4976z^2 + 2.4232z - 0.8796}$$

P2.18 Poles outside the unit circle.

$$G(z) = \frac{z^2 + 0.5657z + 0.16}{z^4 - 1.2178z^3 + 0.3902z^2 + 1.2129z - 0.3125}$$

The zeros of $G(z)$ do not affect the system's stability. They do affect the amplitudes of the mode functions in the system's response, however, and they can block the transmission of input signals. To assess the zeros' impact on the amplitude of the mode functions, a partial-fraction expansion is performed on the z-transform of the output signal, and the residue for each mode function is computed. Certain properties of zeros can be readily illustrated in the discrete-time domain, as shown in Examples 2.10 and 2.11.

☐ EXAMPLE 2.10 *Blocking Zeros*

Build a ZPK object for the system whose transfer function is

$$G(z) = \frac{(z+0.3)(z-\epsilon^{j0.5})(z-\epsilon^{-j0.5})}{(z-0.9)(z-0.7)^2(z+0.5)}$$

Then extract the numerator and denominator polynomials and use the `roots` command to calculate their zeros, and verify that they agree with $G(z)$. Finally, using a sampling period of $T_s = 1$ s, find and plot the system's output, $y(k)$, to the input, $u(k) = \sin(0.5k)$, for $0 \leq k \leq 50$. Explain why the output does not contain any sinusoidal component even though the input is sinusoidal.

Solution

In Script 2.10, we define `zz` to be the three zeros of $G(z)$, `pp` to be the four poles, `gn` to be the gain, and `Ts` to be the sampling period. Then we use them to create `G` as a ZPK object.[2] Note that the gain and the sampling period are the third and fourth input arguments of the `zpk` command, respectively. After drawing the pole-zero plot, we use the `tfdata` command with the `'v'` option to extract the numerator and denominator polynomials from `G`, yielding `numG = [0 1.0 -1.4552 0.4735 0.30]` and `denG = [1.0 -1.80 0.60 0.434 -0.2205]`. Finally, we use the `roots` command on the numerator polynomial, `numG`, to calculate the zeros of $G(z)$, which agree with the values entered as `zz`. For complex discrete-time system poles and zeros, we convert them to the polar form as well, so their magnitudes are displayed. This conversion can be achieved by applying the RPI function `xy2p` to the complex poles or zeros.

To obtain the time response, we first generate the discrete-time vector `k = [0:50]` and the input vector `u = sin(0.5*k)`. Then the `lsim(G,u)` command is used to simulate the time response $y(k)$, which is shown in Figure 2.9, along with the sinusoidal input. Note that the output consists solely of the modes of $G(z)$ and does not exhibit any sinusoidal component as a result of the input $u(k)$.

To understand this phenomenon, we obtain the z-transform of $u(k)$ as

$$U(z) = \mathcal{Z}\{u(k)\} = \mathcal{Z}\{\sin(0.5k)\} = \frac{z\sin(0.5)}{z^2 - 2z\cos(0.5) + 1}$$

[2] MATLAB will accept either `i` or `j` as $\sqrt{-1}$. Although we use j in the text, we use `i` in the scripts and code fragments because MATLAB uses `i` when it displays complex numbers.

Using the fact that $\epsilon^{j0.5} = \cos(0.5) + j\sin(0.5)$, we obtain the z-transform of the output $y(k)$ as

$$Y(z) = G(z)U(z) = \frac{(z+0.3)(z^2 - 2z\cos(0.5) + 1)}{(z-0.9)(z-0.7)^2(z+0.5)} \cdot \frac{z\sin(0.5)}{(z^2 - 2z\cos(0.5) + 1)}$$

Note that the zeros of $G(z)$ at $z = \epsilon^{\pm j0.5}$ cancel the poles of $U(z)$. The MATLAB commands in Script 2.10 compute the discrete-time response and make the plots.

MATLAB Script

```
% Script 2.10:  blocking zeros
zz = [-0.3; exp(0.5i); exp(-0.5i)]    % zeros of G(z)
pp = [0.9; 0.7; 0.7; -0.5]            % poles of G(z)
gn = 1, Ts = 1                        % unity gain & sampling period
G = zpk(zz,pp,gn,Ts)                  % build G(z) as ZPK object
ucircle, axis equal, hold on          % set up unit circle
pzmap(G), hold off                    % plot poles & zeros
[numG,denG] = tfdata(G,'v')           % zeros determined as roots
zGroots = roots(numG)                 % ..of numerator polynomial
%--- magnitudes and angles of zeros from tfdata command ---
[mag_zGroots,theta_zGroots] = xy2p(zGroots)
zG = tzero(G)                       % zeros determined by 'tzero' command
%--- magnitudes and angles of zeros from tzero command ---
[mag_zG,theta_zG] = xy2p(zG)
k = [0:50];                         % generate discrete time array
u = sin(0.5*k);                     % generate discrete input sequence signal
ylsim = lsim(G,u,k);                % compute system response due to u(k)
stem(k,ylsim,'filled'), grid        % plot response
hold on
plot(k,u,'o'), hold off             % plot sinusoidal input
```

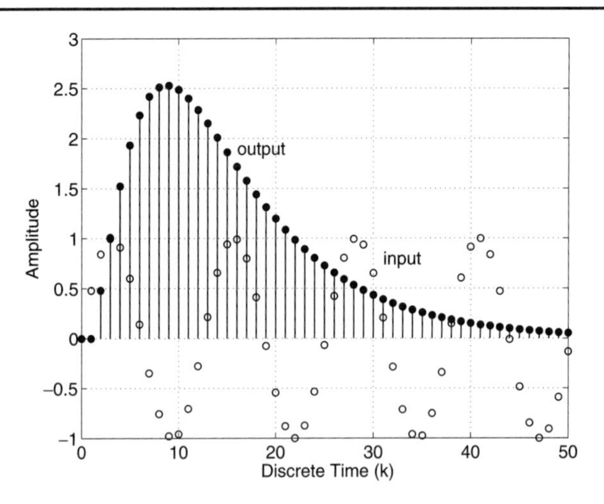

FIGURE 2.9 *Time response for Example 2.10*

WHAT IF? Suppose the frequency of the sinusoidal input is changed to $\omega - 0.2, 0.52$, and 1.2 rad. Simulate the responses of the system in Example 2.10, and observe that there will be sinusoidal components in $y(k)$. Also note that the sinusoidal component in $y(k)$ will be the smallest for $\omega = 0.52$ rad, because the system attenuates signals with frequencies close to $\omega = 0.5$ rad. ■

□

REINFORCEMENT PROBLEMS

In each of the following problems, find the output $y(k)$ for the transfer function $G(z)$ given the input signal $u(k)$ by defining the discrete-time vector k and the input signal $u(k)$, and using the `lsim` command to determine $y(k)$. Note that the input signal characteristics are not present in the output response. Justify this result by showing that $G(z)$ has one or more zeros that cancel the poles of $U(z)$, the transform of the input.

P2.19 Sinusoidal input sequence.

$$G(z) = \frac{2(z - \epsilon^{j\pi/10})(z - \epsilon^{-j\pi/10})}{(z-0.2)(z-0.4)(z-0.6)(z-0.8)} \quad \text{and}$$

$u(k) = 10\sin(\pi k/10)$ for $0 \leq k \leq 40$.

P2.20 Step response of first-order system.

$$G(z) = \frac{z-1}{z+0.5} \quad \text{and} \quad u(k) = \text{unit-step function}$$

What is the steady-state value of the output? *Note:* Because the input is the unit-step function, the `step` command can be used instead of `lsim`.

□ EXAMPLE 2.11 *Zero Outside the Unit Circle*

Plot the zero and the poles in the z-plane of the discrete-time hydroturbine system (with the gate valve dynamics included), whose transfer function is

$$G(z) = \frac{-0.3354z + 0.3526}{z^2 - 1.724z + 0.7408}$$

Include the unit circle in the plot. Also, compute and plot the response to a unit-step input. Use 0.1 s as the sampling period.

Solution

We build the model as a TF object by using the numerator and denominator polynomial coefficients, `[-0.3354 0.3526]` and `[1 -1.724 0.7408]`, as the first and second arguments of the `tf` command, followed by the sampling period, `Ts = 0.1`. We can then find the zero and the poles from the TF object by entering the command `[zG,pG,kG] = zpkdata(G,'v')`. Doing this, we see that the zero is $z = 1.0513$, which is outside the unit circle. The poles are $z = 0.8146$ and 0.9094, which means that the system is stable.

Systems with zeros outside the unit circle are known as *non-minimum phase systems.* The step response is obtained from `step(G)` and is shown in Figure 2.10. Note that even though the input is a positive step, the response initially goes negative. Then it increases and eventually reaches a positive steady-state value equal to the DC gain of 1.0. This behavior is characteristic of non-minimum phase systems with a single zero outside the unit circle.

MATLAB Script

```
% Script 2.11:  zero outside unit circle
% create TF object with Ts = 0.1 s
G = tf([-0.3354  0.3526],[1  -1.724  0.7408],0.1)
[zG,pG,kG] = zpkdata(G,'v')
ucircle, axis equal, hold on          % draw unit circle & set axes
pzmap(G), hold off                    % plot poles and zero in z-plane
[y,k] = step(G);                      % compute step response of Gtf(z)
stem(k,y,'filled',':')                % plot step response
```

WHAT IF? Repeat Example 2.11 with the numerator of $G(z)$ changed to $-z + 1.5$, which places the zero farther away from the unit circle at $z = 1.5$. In the resulting step-response plot, note the smaller initial decrease in the response. This illustrates that moving the zero which is outside the unit circle farther away from the circle reduces that zero's effect on the system response. ■

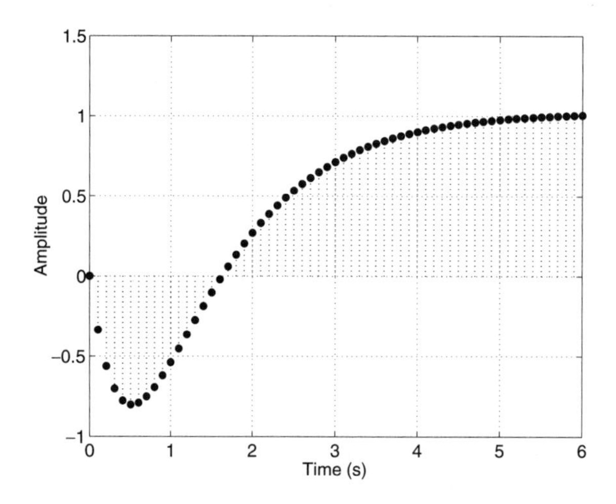

FIGURE 2.10 *Time response for Example 2.11*

□

BUILDING MULTIPLE-INPUT, MULTIPLE-OUTPUT SYSTEMS

So far, the MATLAB activities in this chapter have dealt only with single-input, single-output systems. Practical systems, such as those contained in Appendix A, typically have multiple inputs and outputs. Using LTI objects, the creation of multiple input-output systems from single input-output systems is quite straightforward in MATLAB. A single-input, two-output system whose individual transfer functions are given by $Y_1(z) = G_1(z)U(z)$ and $Y_2(z) = G_2(z)U(z)$ is expressed as

$$\begin{bmatrix} Y_1(z) \\ Y_2(z) \end{bmatrix} = \begin{bmatrix} G_1(z) \\ G_2(z) \end{bmatrix} U(z) = G_c(z)U(z) \tag{2.9}$$

In MATLAB, given the LTI objects `G1z` and `G2z`, the single-input, two-output system is given by `Gcz = [G1z;G2z]`, analogous to the matrix notation of (2.9). Note that this operation is permissible only if the dimensions of the subsystems `G1z` and `G2z` are compatible and the sampling periods are the same.

Similarly, a two-input, one-output system whose individual transfer function is given by $Y(z) = G_1(z)U_1(z) + G_2(z)U_2(z)$ is expressed as

$$Y(z) = \begin{bmatrix} G_1(z) & G_2(z) \end{bmatrix} U(z) = G_r(z)U(z) \tag{2.10}$$

where

$$U(z) = \begin{bmatrix} U_1(z) \\ U_2(z) \end{bmatrix}$$

In MATLAB, given the LTI objects `G1z` and `G2z`, the two-input, one-output system is given by `Grz = [G1z G2z]`, analogous to the matrix notation of (2.10).

A subsystem of a multiple-input, multiple-output system can be obtained using the appropriate indices. For example, to find the transfer function from the first and third inputs to the second output of a multiple-input, multiple-output system object `Gz`, we apply the command `Gsubz = Gz([2],[1 3])`.

In the following example, we build a single-input, two-output system to illustrate the use of MATLAB for combining subsystems.

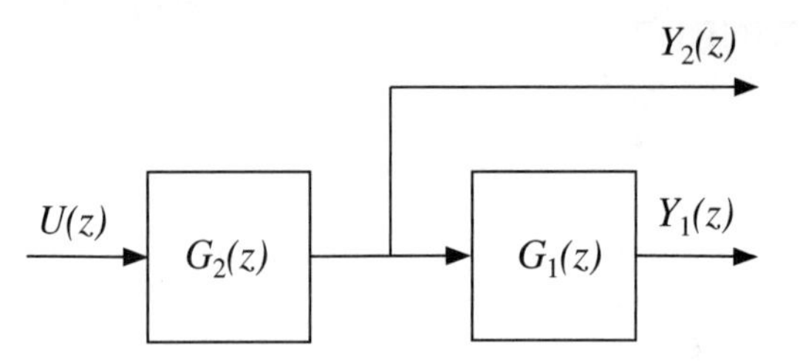

FIGURE 2.11 *Block diagram of the 1-input, 2-output system for Example 2.12*

☐ EXAMPLE 2.12 *A System with 1 Input and 2 Outputs*

A discrete-time model[3] shown in Figure 2.11 has a single input, $u(k)$, and two outputs, $y_1(k)$ and $y_2(k)$. The individual transfer functions are $G_2(z) = Y_2(z)/U(z)$ and $G_1(z) = Y_1(z)/Y_2(z)$, where

$$\begin{aligned} G_1(z) &= \frac{0.06059}{z - 0.9394} \\ G_2(z) &= \frac{0.2212}{z - 0.7788} \end{aligned}$$

Given the model as an LTI object, determine the two transfer functions and their respective zeros, poles, and gain. Assume a sampling period of $T_s = 0.05$ s. Also, simulate the unit-step response, plotting both outputs. Extract the transfer function $G_1(z)$ from the composite system.

Solution

Because we do not cover multiblock system models until the next chapter, we just accept the fact that the first four lines of code in Script 2.12 build a one-input, two-output discrete-time model as the LTI object `G` in TF form.

Using this model, we can obtain its zeros, poles, and gains by using the command `[zz,pp,kk] = zpkdata(G)`. The results are displayed as

```
zz =
      []
      []
pp =
     [2x1 double]
     [     0.7788]
kk =
     0.0134
     0.2212
```

Note that `zz`, `pp`, and `kk` each contain two elements, one for each of the two transfer functions. The two `[]` symbols for `zz` indicate that neither transfer function has any zeros in the finite z-plane. The output for `pp` shows that the poles of $G(z)$ are located in a 2×1 cell array of double-precision numbers. The second transfer function, $G_2(z)$, has a single pole at $z = 0.7788$. To display the contents of the cell array, we enter the command `pp{:}`, which produces

[3]The structure of the model is similar to that of the ball and beam system described in Appendix A.

```
ans =
    0.9394
    0.7788
ans =
    0.7788
```

The interpretation of this output is that $G(z)$ has two poles, one at $z = 0.9394$ and the other at $z = 0.7788$, and $G_2(z)$ has a single pole at $z = 0.7788$.

The calculation of the system's step response is straightforward, requiring only the command `y = step(G,dtime)`. The result `y` will be a matrix, however, having one row for each time point and two columns. The first column of `y` can be designated as `y(:,1)` and contains the values of $y_1(k)$. Similarly, the second column of `y` can be written as `y(:,2)` and contains the values of $y_2(k)$.

Note that the `plot` command in the script has been written so as to plot both outputs, with different symbols, as shown in Figure 2.12. We see that $y_2(k)$ is the response of a fast first-order system, whereas $y_1(k)$ is a somewhat slower second-order response. By studying the poles of the two transfer functions, we can see that these are reasonable results for the system under consideration.

MATLAB Script

```
% Script 2.12:  System with 1 input and 2 outputs
Ts = 0.05
G2 = tf(0.2212,[1  -0.7788],Ts)     % faster system, output = y2
G1 = tf(0.06059,[1  -0.9394],Ts)    % slower system, output = y1
G =  [G1; 1]*G2                     % build 1-input/2-output system
[zz,pp,kk] = zpkdata(G)             % get its zeros, poles, & gains
pp{:}                               % display contents of cell array pp
dtime = 0:Ts:4;
y = step(G,dtime);                  % compute step responses of y1 & y2
plot(dtime,y(:,1),'o',dtime,y(:,2),'x') % plot y1 & y2
```

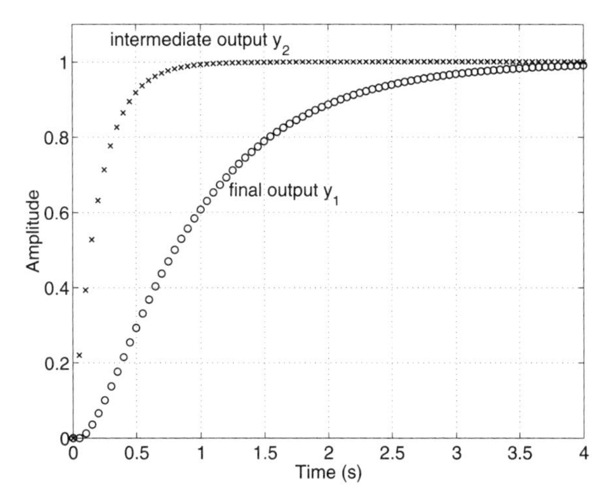

FIGURE 2.12 *Step responses of 1-input, 2-output system in Example 2.12*

□

EXPLORATORY PROBLEMS

EP2.1. Interactive exploration. The file `done_blk.m` is a menu-driven program written in MATLAB that assists the user in building a discrete-time, one-block system model in any of several forms, TF, ZPK, or SS (the state-space form discussed in Chapter 4), and in converting the model between these forms. Once a model has been specified in either TF or ZPK form, the user can select various types of analysis from a menu, including the plotting of step and impulse responses and poles and zeros in the z-plane. It is also possible to obtain time responses to arbitrary inputs. The default system model is that from Example 2.3. The user can enter new values at the prompt, however, and the most recently used values are remembered. Thus they can be reused merely by hitting the Enter key, or they can be modified easily.

You are encouraged to use this program by entering the command `done_blk` and letting your curiosity direct you as to where to go. Some suggestions of topics that have been investigated in this chapter's examples are repeated poles, the effect of a zero outside the unit circle on the step response, the relationship of the angle of a complex pole and the frequency of oscillation, and the effect of the magnitude of a pole on the rate of decay (or growth) of the envelope of the response.

COMPREHENSIVE PROBLEMS

The following problems differ from the others in two ways: (i) they all relate to either laboratory experiments or real-world system models described in Appendix A, and (ii) you are asked to apply the MATLAB tools that have been introduced in the chapter. Specifically, for the designated system model, you should run the appropriate M-file to generate the requested transfer function in TF form using a pair of row vectors that contain the transfer function's coefficients. These coefficient vectors are always named `dnum` and `dden`, for the numerator and denominator polynomials, respectively. The sampling period is given in the variable `Ts`.

Because you will be working with a model in TF form in each case, you can use the `zpkdata` command to see the zeros and poles of the transfer function $G(z)$ or use the `residue` command to obtain its partial-fraction expansion. The `step` and `impulse` commands will generate and plot the step and impulse responses. Keep in mind that if you use these commands without specifying a time vector, MATLAB will select one. Depending on the particular system with which you are dealing, the time interval MATLAB uses may or may not make sense. So be prepared to define your own time vector and include it as an optional argument.

In the problem statements, we identify the system model to be considered and specify one or more of its transfer functions for analysis. You should feel free to use any of the other transfer functions we do not specifically mention.

CP2.1 **Ball and beam system.** Run the MATLAB file `dbbeam.m` to develop the two transfer functions of the discrete-time ball and beam system with a sampling period of $T_s = 0.02$ s. The input is the voltage applied to the motor that rotates the beam. The two outputs for which transfer functions are developed in `dbbeam.m` are the angle of the wheel (θ) in radians and the position of the ball (ξ) in meters. What are the orders of these two transfer functions? Examine the poles and zeros of each transfer function and comment on their similarities and differences. Also do a partial-fraction expansion and relate its results to the poles and zeros of the two transfer functions. Examine the time response the model produces when the voltage applied to the motor is a positive pulse of unity magnitude for 0.1 s followed by a negative pulse of the same magnitude for 0.1 s. Repeat the simulation with the positive and negative pulses separated in time. Compare the various time responses and provide an explanation for their differences.

CP2.2 **Inverted pendulum.** Run the file `dstick.m` to develop the two transfer functions of the discrete-time inverted pendulum system with a sampling period of $T_s = 0.01$ s. The input u is the voltage applied to the motor that drives the cart. The two outputs for which transfer functions are developed in `dstick.m` are the angle of the pendulum (θ) and the position of the cart (x). Two other outputs for which transfer functions could be computed are the angular velocity of the pendulum (ω) and the velocity of the cart (v). Examine the poles and zeros of each transfer function and comment on their similarities and differences. Also do a partial-fraction expansion and relate its results to the poles and zeros. Try to lend some physical significance to the poles. For example, why is there a pole outside the unit circle and how does its value relate to the time responses?

Examine the time responses produced by the model when the inputs are (i) a step and (ii) an impulse in the voltage applied to the motor that drives the cart. Using the sign conventions shown in Figure A.5, give physical arguments that the directions of the translational and rotational motions are correct.

CP2.3 **Electric power generation system.** Run the MATLAB `dpower.m` file to obtain the transfer functions from the one input (field voltage V_{field}) to each of the three outputs (terminal voltage V_{term}, machine speed ω, and electrical power P). The sampling period is $T_s = 0.01$ s. Verify that the poles of all three transfer functions are identical and have the values $z = 0.9989, 0.9952\epsilon^{\pm j0.0933}$, 0.9697, 0.7652, 0.7019, and 0.3175. Calculate the zeros of each of the transfer functions and show that they are different from one another. Use the `pzmap` command to generate pole-zero plots for each of the transfer functions. Also do a partial-fraction expansion for each transfer function and examine the relative weightings of the mode functions.

Using the `step` command, plot the responses of each of the outputs to a step function of amplitude 0.05 in V_{field}. Then relate the character of the response to the dominant poles—those closest to the unit circle—namely, $z = 0.9989$ and the complex pair at $z = 0.9909 \pm j0.0928 = 0.9952\epsilon^{\pm j0.0933}$.

Note that the real pole at $z = 0.9989$ results in a time constant of $\tau = T_s/\ln(1/z) = 0.01/0.0011 = 9.09$ s. By examining the three step responses, you should see that this slowly decaying mode appears only in the terminal voltage response. This should convince you that the residues and pole-zero plots substantiate what the responses show. The pair of complex poles at $z = 0.9952\epsilon^{\pm j0.0933} = r\epsilon^{\pm j\theta}$ produce lightly damped oscillations having a period of $T_p = (2\pi/\theta)T_s = (2\pi/0.0933) \times 0.01 = 0.673$ s. Verify this result by plotting the oscillatory responses with a time scale that allows you to measure the period of the oscillation. You should see that it takes just over 2 seconds for three cycles of the oscillation, which agrees with the value of the period computed from the pole.

CP2.4 Hydroturbine system. In Example 2.11, we observed that the step response of this second-order, non-minimum phase model of the hydroturbine system initially goes negative before becoming positive. To see mathematically why this happens, do a partial-fraction expansion and observe that the residue at one of the two poles is negative. Use the RPI function `rpole2k` to compute and plot the parts of the step response due to each of the poles.

SUMMARY

We have used MATLAB to build transfer-function models of single-block systems in several forms: (i) as a ratio of polynomials, referred to here as the TF form, (ii) as lists of the zeros, poles, and gain, referred to as the ZPK form, and (iii) as a partial-fraction expansion involving the sum of first-order terms. We have also shown how MATLAB and its Control System Toolbox can be used to perform a variety of analytical operations on these transfer-function models. Specifically, we have computed and plotted the system's responses to step functions, impulses, and general inputs. We have also shown how to obtain the contributions to the total response of individual poles, both real and complex, and have looked at the effects of having repeated poles. Finally, we have demonstrated the effects of the transfer function's poles on the stability of the system and the effect of its zeros on the response.

With these fundamentals established, we are now prepared to consider the construction and analysis of system models that involve more than one block. These topics will be the subject of Chapter 3.

ANSWERS

P2.1 Zeros are $z = 0.40, 0.60$; poles are $z = 0.1505, 0.8153\epsilon^{\pm j0.9864}$; gain = 1.250.

P2.2 Zeros are $z = -0.5271, 0.3619\epsilon^{\pm j0.5915}$; poles are $z = 0.2872\epsilon^{\pm j2.2079}$, $0.7786\epsilon^{\pm j0.4931}$; gain = 0.840.

P2.3 $G(z) = (150z^2 - 30z - 12)/(z^4 + 0.7z^3 + 0.3325z^2 - 0.4237z - 0.1462)$

P2.4 $G(z) = (5z^3 + 5.5z^2 + 2.2z + 0.3)/(z^4 - 0.8z^3 - 0.21z^2 + 0.288z - 0.054)$

P2.5 Poles of $G(z)$ are $z = -0.2592, 0.6211\epsilon^{\pm j1.9335}$, and the residues of the poles are $-2.5999, 1.4785\epsilon^{\pm j2.5178}$, respectively. The residue at $z = 0$ is 5.0. The impulse response has a maximum value of 0.960 at $k = 2$ and a minimum value of -0.3096 at $k = 4$.

P2.6 Poles of $G(z)$ are $z = -0.2501, 0.3502, 0.650\epsilon^{\pm j0.3949}$, and the residues of the poles are $3.5526, 25.7766, 8.3562\epsilon^{\pm j2.8125}$, respectively. The residue at $z = 0$ is -13.5135. The impulse response peaks at 1.4294 for $k = 5$ and then decays to zero as $k \to \infty$, but it does not do so monotonically.

P2.9

$$G(z) = \frac{1.0468z}{z - 0.60} + \frac{0.1818z}{(z + 0.50)^2} + \frac{1.6198z}{z + 0.50} - 2.6667$$

P2.10

$$G(z) = \frac{3.7755z}{(z + 0.5)^2} + \frac{15.4810z}{z + 0.5} - \frac{2.9388z}{(z - 0.20)^2} + \frac{34.5190z}{z - 0.20} - 50.0$$

P2.11 Step response reaches a maximum value of 1.160 at $k = 2$ and approaches a steady-state value of $y_s(\infty) = 1.80/2.30 = 0.7826$.

P2.12 Step response reaches a maximum value of 6.7966 at $k = 11$ and approaches a steady-state value of $y_s(\infty) = 1.20/0.1808 = 6.6372$.

P2.13 The response reaches a maximum of 3.2692 at $t = 2.4$ s and settles to within 1 percent of its final value of 3.1304 by $t = 3.0$ s.

P2.14 The response reaches a maximum of 7.6878 at $t = 1.4$ s and a minimum of -7.1361 at $t = 2.6$ s. It settles to within ± 0.05 of its final value of 0 by $t = 5.0$ s.

P2.15 Poles are $z = 0.9367\epsilon^{\pm j2.6884}, 0.9549\epsilon^{\pm j1.0894}$. The system is stable.

P2.16 Poles are $-1, 0.50, 0.2828\epsilon^{\pm j0.7854}$. The system is marginally stable because of the pole at $z = -1$. The plot of the impulse shows an undamped oscillation with an amplitude of 0.5856, a period of $k = 2$, and an average value of 0.

P2.17 Poles are $z = 0.8796, 1.0002\epsilon^{\pm j0.6283}$. The steady-state part of the impulse response is a sinusoid whose period and amplitude are approximately 10 s and 3.7, respectively.

P2.18 Poles are $z = -0.80, 0.250, 1.25\epsilon^{\pm j0.7854}$. The system is unstable, and the impulse response contains a sinusoid whose amplitude grows *exponentially* with time.

P2.19 The response rises rapidly from 0 to a maximum of 17.305 at $k = 6$ and then decays monotonically, coming within 0.10 of its final value of 0 by $k = 33$. The sinusoidal input is not present in the response.

P2.20 The response starts at 1.0, goes to -0.5, and continues to oscillate as its amplitude decays to zero. The step input is not present in the response.

3 *Building and Analyzing Multiblock Models*

PREVIEW

A discrete-time control system is usually described in terms of an interconnection of components, where each component is described by a set of difference equations or a transfer function. Based on the interconnection information, often given in the form of a block diagram, a designer builds the control system model from the descriptions of the component models. In this chapter we discuss three basic model interconnections—series, parallel, and feedback—and show how to interconnect models given in transfer-function form using commands from the Control System Toolbox. We also illustrate the property responsible for the fact that some interconnections preserve the zeros of the component models whereas others preserve the poles of the component models. Interconnections for models given in state-space form will be discussed in Chapter 4.

SERIES CONNECTIONS

Figure 3.1 shows two discrete-time systems having the same sampling period (T_s) that are connected in series (also referred to as a *cascade connection*). Provided that no loading effects are present, the transfer function of the resulting system is the product of the individual transfer functions, namely $T(z) = G_2(z)G_1(z)$. Note that we have placed the transfer function of the left-hand block *after* that of the right-hand block. The reason for doing this is evident if we define the output of $G_1(z)$ to be $V(z) = G_1(z)U(z)$ and express the output of the second block as $Y(z) = G_2(z)V(z)$. Using the first equation to substitute for $V(z)$ in the second equation, we see that the proper expression for the transform of the output is

$$Y(z) = G_2(z)G_1(z)U(z)$$

which says that the transfer function of the series connection is

$$T(z) = G_2(z)G_1(z) \tag{3.1}$$

For single-input/single-output (SISO) systems in either TF or ZPK form, or a mixture of the two forms, the results will be the same for $G_2(z)G_1(z)$ or $G_1(z)G_2(z)$. If, as will be done in Chapter 4, the state-space form is used, however, the internal representations, in terms of the state-space matrices, will be different. More important, for systems that are not SISO, the operation $G_1(z)G_2(z)$ may not even be defined and will most likely give incorrect results if it is defined. Thus, we use the expression in (3.1) for the series connection depicted in Figure 3.1.

It follows that the poles of $T(z)$ are the combined poles of $G_1(z)$ and $G_2(z)$ and that the zeros of $T(z)$ are the combined zeros of $G_1(z)$ and $G_2(z)$. Should a zero of $G_2(z)$ coincide with a pole of $G_1(z)$ or a zero of $G_1(z)$ be the same as a pole of $G_2(z)$, the pole-zero pair would cancel, and neither would show up in $T(z)$. Should both $G_1(z)$ and $G_2(z)$ have an identical pole or zero, the multiplicity of that pole or zero would be increased accordingly.

Because the poles of $T(z)$ are the combined poles of $G_1(z)$ and $G_2(z)$, the series connection $T(z)$ has the same stability property as $G_1(z)$ and $G_2(z)$, provided that they do not have common poles on the unit circle. The stability property of $T(z)$ should be determined before any pole-zero cancellation occurs. For example, if $G_1(z)$ has a pole at $z = 1$ and $G_2(z)$ has a zero at $z = 1$, $T(z)$ is (internally) marginally stable even though the symptoms are not observed at the interconnection's output.

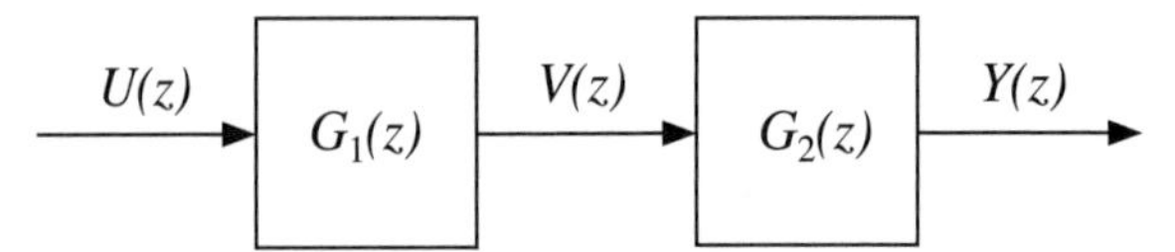

FIGURE 3.1 *Two systems connected in series*

Expressing linear time-invariant systems as LTI objects has the advantage of system interconnections becoming more transparent. For example, the series connection of the two systems shown in Figure 3.1, $G_1(z)$ and $G_2(z)$, is mathematically expressed as $G(z) = G_2(z) \times G_1(z)$. In MATLAB, the series connection is obtained by the command `G = G2*G1`, where `G1` and `G2` are LTI objects in any form. When operating on two LTI objects, the symbol `*` means *series* connection. Because the interpretation of `*` is based on the properties of its operands, it is an example of an *overloaded operator.*

When connecting two or more systems, you need to consider what the type of the overall system will be (e.g., TF or ZPK). You can determine this using a set of precedence rules established in the Control System Toolbox, which are based on the mathematical properties of the different forms. For now, you only need to know that if all the subsystems are in TF form, the resulting system will be in TF form. If any one of the subsystems is in ZPK form, however, the overall system will be in ZPK form. This is because the ZPK form has better mathematical properties than the TF form. We will revisit these precedence rules in Chapter 4 when the state-space form is introduced.

☐ EXAMPLE 3.1 *Series Connection of Two Systems*

Create the series connection, denoted by the transfer function $T(z)$, of the two systems whose transfer functions are $G_1(z)$ and $G_2(z)$, where

$$G_1(z) = \frac{0.02(2z + 1.4)}{z^2 - 1.7z + 0.72}$$

and $G_2(z)$ has a zero at $z = -0.2$, poles at $z = 0.9\epsilon^{\pm j\pi/5}$ and -0.8, and a gain of 0.5. Both systems are sampled at $T_s = 1$ s. Create both the TF and ZPK forms of the series connection. Then plot the unit-step responses over the interval $0 \leq k \leq 50$ and show that they are the same. Finally, determine the stability of the series connection and relate its zeros, poles, and gain to those of $G_1(z)$ and $G_2(z)$.

Solution

The MATLAB commands in Script 3.1 start by building $G_1(z)$ as a TF object and determining its zeros, poles, and gain. Then $G_2(z)$ is built in ZPK form, as its zeros, poles, and gain are known. Next the `*` operator is used to connect the two systems, resulting in the transfer function $T_{\text{zpk}}(z)$. Because of the precedence rules, this system will automatically be in ZPK form, with its transfer function displayed in factored form. To see the zeros, poles, and gain listed, we use the `zpkdata` command with the `'v'` option. To create the TF form of the series connection, we use the `tf` command, with `Tzpk` as its argument. In this case, the transfer function is displayed in unfactored form.

To show that the two forms (ZPK and TF) have the same unit-step responses, we use the `step` command with each form. First, we plot the response of the ZPK system with open circles and use the `hold on` command to overplot the response of the TF system with the `+` symbol. In the resulting plot, shown in Figure 3.2, all of the `+` symbols fall inside the circles of the stem plot, proving that the responses of the two systems are indistinguishable, at least to the resolution of the plot. To check this assertion in more quantitative terms, we can compute the maximum absolute value of the difference between the two response vectors, **yzpk** and **ytf**, and show that it is a small number. To do this, we enter the MATLAB expression **max(abs(yzpk - ytf))**.

MATLAB Script

```
% Script 3.1:  Series connection of two blocks
Ts = 1
G1 = tf(0.02*[2 1.4],[1 -1.7 0.72],Ts)      % G1(z) in TF form
[zG1,pG1,kG1] = zpkdata(G1,'v')             % zeros, poles, & gain of G1
pp = 0.9*exp(j*pi/5)
pG2 = [pp; conj(pp); -0.8]           % poles of G2 as column vector
G2 = zpk(-0.2,pG2,0.5,Ts)            % G2 in ZPK form
Tzpk = G2*G1                         % series combination in ZPK form
[zT,pT,kT] = zpkdata(Tzpk,'v')       % zeros, poles, & gain from ZPK form
Ttf = tf(Tzpk)                       % series combination in TF form
dt=[0:Ts:50]';                       % discrete time sequence
yzpk = step(Tzpk,dt);                % compute step response from ZPK form
ytf = step(Ttf,dt);                  % compute step response from TF form
%------ plot both responses on same axes --------------
stem(dt,yzpk,':'),grid,hold on       % plot ZPK response with dotted stems
plot(dt,ytf,'+'),hold off            % plot TF form with '+' symbol
```

The transfer function of the series combination is

$$T(z) = \frac{0.02z^2 + 0.018z + 0.0028}{z^5 - 2.356z^4 + 1.481z^3 + 0.779z^2 - 1.357z + 0.4666} \tag{3.2}$$

which has zeros at $z = -0.7$ and -0.2, poles at $z = -0.8, 0.9\epsilon^{\pm j0.6283}, 0.8$, and 0.9, and a gain of 0.02. Because all of the poles of $G_1(z)$ lie inside the unit circle, the series combination is stable.

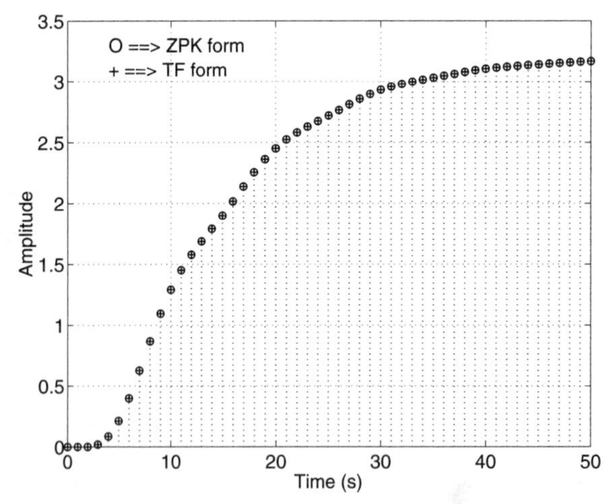

FIGURE 3.2 *Step responses for Example 3.1*

We can see that the zeros of $T(z)$ are those of $G_1(z)$ [$z = -0.7$] and $G_2(z)$ [$z = -0.2$]. Likewise, the poles of $T(z)$ are the poles of $G_1(z)$ [$z = 0.8$ and 0.9] and the poles of $G_2(z)$ [$0.9\epsilon^{\pm j0.6283}$ and -0.8]. Finally, the gain of $T(z)$ is the product of the gain of $G_1(z)$ and the gain of $G_2(z)$ [$0.04 \times 0.5 = 0.02$].

□

REINFORCEMENT PROBLEMS

Use MATLAB to make series connections of the type shown in Figure 3.1 for the systems whose transfer functions are given below. In each case, create the transfer function $T(z)$ of the combined system using a ratio of polynomials and determine its zeros, poles, and gain. Determine the stability of $T(z)$ and plot the response to the unit step function. Also, identify the sources of the poles and zeros of $T(z)$. The sampling period is unity.

P3.1 Two second-order systems.

$$G_1(z) = \frac{2z^2 - 1.414z + 0.5}{z^2 - 0.7z + 0.49} \quad \text{and} \quad G_2(z) = \frac{z + 0.25}{z^2 - 0.494z + 0.64}$$

P3.2 Third- and first-order systems.

$$G_1(z) = \frac{3(z - 0.4)}{(z + 0.1)(z^2 - 0.4z + 0.29)} \quad \text{and} \quad G_2(z) = \frac{4}{z - 0.8}$$

P3.3 Interchanging two systems. Repeat Example 3.1, but with the order of the two systems reversed, so that $G_2(z)$ precedes $G_1(z)$. Verify that the resulting system has the same transfer function as that in the example.

P3.4 Three systems in series. $T(z) = G_1(z)G_2(z)G_3(z)$ where

$$G_1(z) = \frac{z + 0.1}{z + 0.5}, \quad G_2(z) = \frac{2(z + 1.5)}{0.5z^2 + 0.2z + 0.2}, \quad \text{and}$$

$$G_3(z) = \frac{3z + 1}{z^2 + 0.2z + 0.8}$$

P3.5 Stable zero-pole cancellation.

$$G_1(z) = \frac{2z - 1}{z^2 + 0.5z + 0.125} \quad \text{and} \quad G_2(z) = \frac{3z + 0.5}{(z^2 - 0.25z + 0.125)(z - 0.5)}$$

P3.6 Common zero and pole outside the unit circle.

$$G_1(z) = \frac{z^2 - 0.2z - 0.24}{z^3 + 0.8z^2 - 0.53z - 0.06} \quad \text{and} \quad G_2(z) = \frac{z + 1.2}{0.64z^2 + 0.01}$$

Two single-input/single-output systems, $H_1(z)$ and $H_2(z)$, having a common sampling period, T_s, can be connected in parallel by joining their inputs and summing their outputs, as indicated in Figure 3.3. The transfer function of the resulting system is

$$T(z) = H_1(z) + H_2(z)$$

which will have poles that are the union of the poles of $H_1(z)$ and $H_2(z)$. The zeros of $T(z)$ will differ from the zeros of $H_1(z)$ and $H_2(z)$, however. Because the poles of $T(z)$ are the union of the poles of $H_1(z)$ and $H_2(z)$, the parallel connection $T(z)$ has the same stability property as $H_1(z)$ and $H_2(z)$. This stability is not affected by $H_1(z)$ and $H_2(z)$ having common poles on the unit circle, as in the series connection. Should a zero of $T(z)$ be the same as a pole of $H_1(z)$ or $H_2(z)$, the pole-zero pair cancels. However, system stability should be determined before any pole-zero cancellations occur.

In MATLAB, the parallel connection of two systems $T(z) = H_1(z) + H_2(z)$ is obtained by the command `T = H1 + H2`. When working with two LTI objects, the symbol + is an overloaded operator indicating *parallel* connection. The following example illustrates this process.

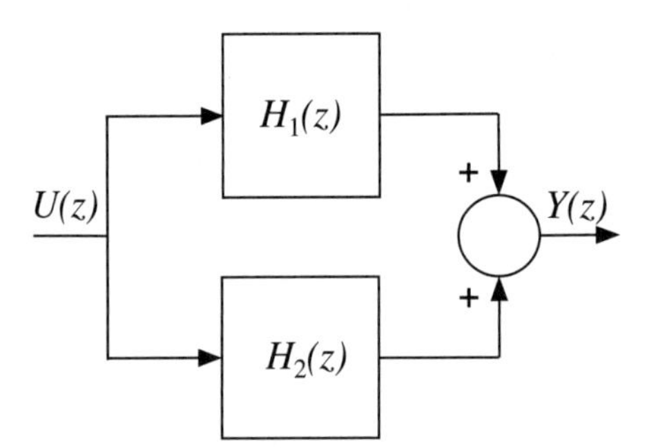

FIGURE 3.3 *Two systems connected in parallel*

☐ EXAMPLE 3.2 *Parallel Connection*

Form the parallel connection of the two transfer functions

$$H_1(z) = \frac{z + 0.5}{z + 0.75} \quad \text{and} \quad H_2(z) = \frac{0.15z}{(z - 0.9)(z - 0.8)(z + 0.8)}$$

both of which have $T_s = 1$ s. Determine the zeros, poles, and gain of $T(z)$ and compare them with the corresponding parameters of $H_1(z)$ and $H_2(z)$. Also determine the stability of $T(z)$ and plot the step response.

Solution A set of MATLAB commands that accomplish the objectives of this example appears in Script 3.2. The values of the zeros, poles, and gain of $H_1(z)$, $H_2(z)$, and $T(z)$ are shown in Table 3.1, where the complex zeros at $z = 0.8295 \pm j0.3022$ have been written in polar form as $0.8829\epsilon^{\pm j0.3494}$.

It is apparent that the poles of $T(z)$ consist of the poles of $H_1(z)$ and $H_2(z)$, but the zeros of $T(z)$ are different from those of $H_1(z)$ and $H_2(z)$, as expected. Because the poles of $H_1(z)$ and $H_2(z)$ are all inside the unit circle, the parallel connection $T(z)$ is stable.

MATLAB Script

```
% Script 3.2:  Parallel connection of two blocks
Ts = 1                               % unity sampling period
H1 = tf([1 0.5],[1 0.75],Ts)         % H1 in TF form
%---- use convolution cmd to multiply 3 polynomials in denominator of H2
denH2 = conv([1 -0.9],conv([1 -0.8],[1 0.8]))
H2 = tf([0.15 0],denH2,Ts)           % H2(z) in TF form
%---- parallel combination will be in TF form -----------
T = H1 + H2
[zT,pT,kT] = zpkdata(T,'v')          % zeros, poles, and gain of T(z)
pH1 = pole(H1)                       % poles of H1
pH2 = pole(H2)                       % poles of H2
[y,k] = step(T);                     % step response of T(z)
stem(k,y,':','filled')               % plot with stem option
```

	$H_1(z)$	$H_2(z)$	$T(z)$
Zeros	-0.5	0	$0.8829\epsilon^{\pm j0.3494}$, $-0.7935, -0.4656$
Poles	-0.75	$0.9,\ 0.8,\ -0.8$	$0.9, 0.8, -0.8, -0.75$

TABLE 3.1 *Zeros and poles of the transfer functions in Example 3.2*

W H A T I F ? Suppose that in Example 3.2 we had formed the difference of the two transfer functions. Use the overloaded operator `-` to create the parallel connection $D(z) = H_2(z) - H_1(z)$. Then determine its zeros and poles, and plot its response to the unit-step function. You should find that although $D(z)$ has the same poles as $T(z)$ in the example, it now has a zero outside the unit circle. Its step response starts out in the negative direction before becoming positive and approaching a positive steady-state value. How would you expect the step response of the system defined by $G(z) = H_1(z) - H_2(z)$ to look? Use MATLAB to check it out. ■

□

REINFORCEMENT PROBLEMS

Use MATLAB to make a parallel connection of the two systems whose transfer functions are given below. In each case, create the transfer function $T(z) = H_1(z) + H_2(z)$ as a ratio of polynomials; determine its zeros, poles, and gain; and plot the response to the unit-step function. Also determine the stability of the interconnected system. The sampling period T_s is unity in all cases.

P3.7 Second- and first-order systems.

$$H_1(z) = \frac{2z - 0.3}{5z^2 + 2z + 2} \quad \text{and} \quad H_2(z) = \frac{3(z + 0.5)}{6z - 1}$$

P3.8 Two second-order systems.

$$H_1(z) = \frac{5z - 4}{2z^2 + 0.4z + 1} \quad \text{and} \quad H_2(z) = \frac{3z^2 + z + 4}{4z^2 + 3z + 2}$$

P3.9 Stable zero-pole cancellation.

$$H_1(z) = \frac{1}{z^2 - 0.1z - 0.02} \quad \text{and} \quad H_2(z) = \frac{2z - 1}{z^2 + 0.1z - 0.06}$$

For what value of z is there a pole-zero cancellation? Is it a stable or unstable cancellation?

☐ EXAMPLE 3.3 *Series/Parallel Connections*

Use the * and + operators to connect the four subsystems in the series/parallel arrangement shown in Figure 3.4 where $G_1(z)$ and $G_2(z)$ are the two blocks from Example 3.1, and $G_3(z)$ and $G_4(z)$ are the two blocks from Example 3.2, with $G_3(z) = H_1(z)$ and $G_4(z) = H_2(z)$. All four subsystems are sampled at $T_s = 1$ s. Verify that the poles of the overall transfer function are the union of the poles of the individual transfer functions. Also verify that the zeros of the overall transfer function do not include the zeros of $G_1(z), G_2(z)$, or $G_3(z)$ but do include the zero of $G_4(z)$. Find and plot the step response of the combined system.

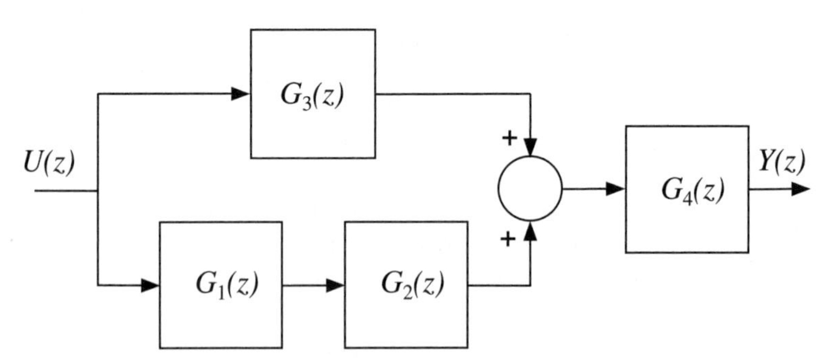

FIGURE 3.4 *A series/parallel connection of four subsystems*

Solution First, we use the definitions of the transfer functions of the individual blocks to create each of them as LTI objects, with `G1`, `G3`, and `G4` being in TF form, `G2` being in ZPK form, and all four blocks having sampling periods of unity. We use the MATLAB convolution function `conv` to form the denominator polynomial of `G4` without creating a separate variable. Once this is done, the series/parallel connection $T(z)$ can be obtained with the single command `T = G4*(G3 + G2*G1)`. Because one of the subsystems, `G2`, has been implemented in ZPK form, the combined system will also be in ZPK form. This can be seen by the form in which $T(z)$ is displayed following the completion of the command—namely,

$$T(z) = \frac{0.15z(z+0.8)(z+0.4997)(z^2-1.774z+0.8961)(z^2-1.382z+0.6571)}{(z-0.9)^2(z-0.8)^2(z+0.8)^2(z+0.75)(z^2-1.456z+0.81)}$$

where the numerator and denominator are factored polynomials in z. Note that the ZPK form displays complex zeros and poles with second-order polynomials.

If the combined system had been in TF form, as would have been the case if all four subsystems were in TF form, the expression for $T(z)$ would have been displayed as a ratio of unfactored polynomials. We can easily determine this form by entering the command `Ttf = tf(T)`. Doing this, we obtain

$$T(z) = \frac{0.15z^7 - 0.2784z^6 + 0.04537z^5 + 0.2309z^4 - 0.1402z^3 - 0.02936z^2 + 0.03531z}{z^9 - 2.506z^8 + 0.519z^7 + 3.751z^6 - 3.215z^5 - 1.23z^4 + 2.429z^3 - 0.4073z^2 - 0.5415z + 0.2016}$$

which is correct, but not particularly informative.

To retrieve the seven zeros and nine poles of the overall system as column vectors, we use `[zT,pT,kT] = zpkdata(T,'v')`. The values are listed in Table 3.2. Then we use the `step` command to compute and plot the step response shown in Figure 3.5. The `dcgain` command computes the steady-state value of the step response as 16.92, which agrees with the figure.

	$G_1(z)$	$G_2(z)$	$G_3(z)$	$G_4(z)$	$T(z)$
Zeros	-0.7	-0.2	-0.5	0	$0, -0.8, -0.4997,$ $0.8106\epsilon^{\pm j0.5507},$ $0.9466\epsilon^{\pm j0.3562}$
Poles	$0.8,$ 0.9	$0.9\epsilon^{\pm j0.6283},$ -0.8	-0.75	$-0.8,$ $0.8,$ 0.9	$0.9\epsilon^{\pm j0.6283},$ $-0.75, -0.8, -0.8,$ $0.8, 0.8, 0.9, 0.9$

TABLE 3.2 *Zeros and poles of the transfer functions in Example 3.3*

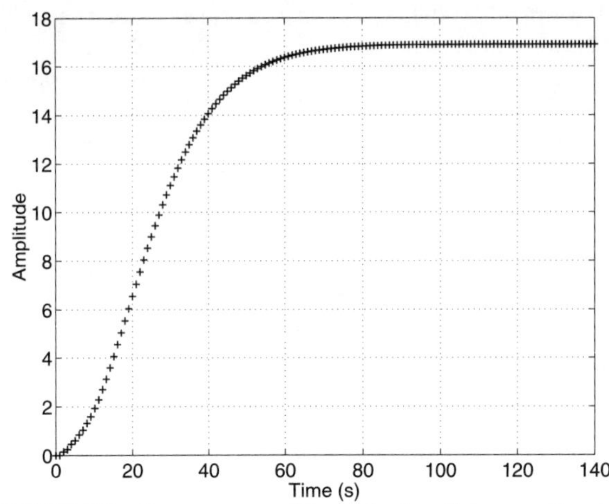

FIGURE 3.5 *Step response of the series/parallel connection in Example 3.3*

The MATLAB commands that create $T(z)$ and carry out the required analysis are given in Script 3.3.

MATLAB Script

```
% Script 3.3:  Series/Parallel connection of four blocks
Ts = 1                                      % unity sampling period
G1 = tf(0.02*[2 1.4],[1 -1.7 0.72],Ts)      % G1 in TF form
pp = 0.9*exp(j*pi/5)
pG2 = [pp; conj(pp); -0.8]                  % poles of G2 as column vector
G2 = zpk(-0.2,pG2,0.5,Ts)                   % G2 in ZPK form
G3 = tf([1 0.5],[1 0.75],Ts)                % G3 in TF form
denG4 = conv([1 -0.9],conv([1 -0.8],[1 0.8]))      % denom of G4
G4 = tf([0.15 0],denG4,Ts)                  % G4 in TF form
T = G4*(G3+G2*G1)                           % overall T(z) in ZPK form
[zT,pT,kT] = zpkdata(T,'v')                 % zeros, poles, gain of T(z)
[resp,k] = step(T);                         % step resp. of T(z)
plot(k,resp,'+'), grid                      % plot response
dcgain(T)                                   % steady-state step response
```

By using the `zpkdata` command on the LTI object `T`, we obtain the zero and pole entries in Table 3.2.

Examination of the entries in Table 3.2 shows that each of the nine poles of $T(z)$ appears as a pole in one of the four subsystems. Also, of the seven zeros of $T(z)$, only the one at $z = 0$ appears as a zero in any of the subsystems. That this is the zero of $G_4(z)$ is due to the fact that the combination of the other three subsystems is in series with $G_4(z)$.

□

REINFORCEMENT PROBLEMS

Use MATLAB to make a series/parallel connection, according to the block diagram given in Figure 3.4, of the four systems whose transfer functions are given below. In each case, create the transfer function of the combined system in TF form; determine its zeros, poles, and gain; and plot the response to the unit-step function. Also determine the stability of the interconnected system. The sampling period is 1 s.

P3.10 Series/parallel connection.

$$G_1(z) = \frac{12z^2 + 5z + 1}{24z^2 + 15z + 6}, \quad G_2(z) = \frac{4(z + 0.9)}{(z + 0.2)(z - 0.8)},$$

$$G_3(z) = \frac{5}{z - 0.5}, \quad \text{and} \quad G_4(z) = \frac{3z + 2}{4z^2 - 3}$$

P3.11 Another series/parallel connection.

$$G_1(z) = \frac{6}{z^2 + z + 1}, \quad G_2(z) = \frac{4z + 1}{16z - 1},$$

$$G_3(z) = \frac{7z + 0.5}{10z + 3}, \quad \text{and} \quad G_4(z) = \frac{4}{z^2 - 0.3z + 0.81}$$

FEEDBACK CONNECTIONS

Figure 3.6 shows two discrete-time systems having a common sampling period, T_s, and connected in a feedback configuration with a negative sign associated with the feedback signal where it enters the summing junction. We refer to $G(z)$ as the forward transfer function and to $H(z)$ as the feedback transfer function. The input and output signals of $G(z)$ are $u(k)$ and $y(k)$, respectively, and the reference input is $r(k)$. The

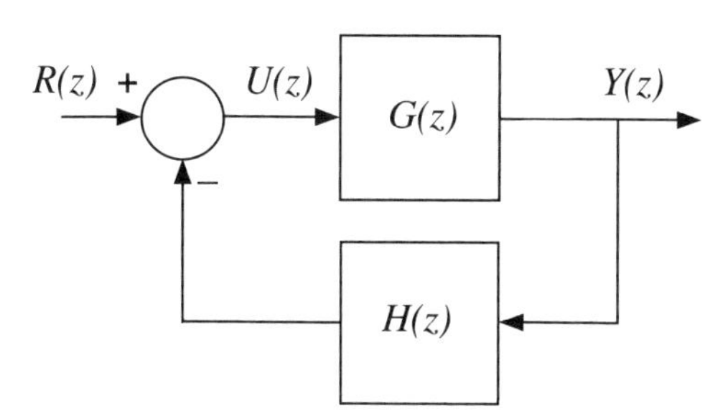

FIGURE 3.6 *A feedback connection of two systems*

transfer function of the complete system from $R(z)$ to $Y(z)$, denoted by $T(z)$, with the negative feedback as shown in the figure, is

$$T(z) = \frac{G(z)}{1 + G(z)H(z)}$$

For systems of this type, the feedback system $T(z)$ can be implemented using the Control System Toolbox `feedback` command, which accepts LTI objects as input arguments. For example, the connection of $G(z)$ and $H(z)$ as shown in Figure 3.6 is obtained with the command `T = feedback(G,H)`, where negative feedback is assumed. If the feedback signal enters the summing junction with a + sign, a third argument of `+1` is included. For unity feedback, the command would be `T = feedback(G,1)`, where the second argument, `1`, represents a system having unity gain.

It can be shown that the closed-loop transfer function $T(z) = Y(z)/R(z)$ will have zeros that are the zeros of $G(z)$ and the poles of $H(z)$. The closed-loop poles will differ from the poles of $G(z)$ and $H(z)$.

☐ EXAMPLE 3.4 *Feedback Connection*

Use MATLAB to create the negative feedback system shown in Figure 3.6 where

$$G(z) = \frac{0.2(z - 0.85)}{(z - 0.9\epsilon^{j\pi/4})(z - 0.9\epsilon^{-j\pi/4})}$$

and

$$H(z) = \frac{z - 0.3}{(z - 0.8)(z - 0.1)}$$

and both sampling periods are unity. Find the closed-loop transfer function as a ratio of polynomials and determine its zeros, poles, and gain. Compare the poles of the closed-loop transfer function $T(z)$ to the poles of $G(z)$ and $H(z)$. Determine the stability of $T(z)$. Also verify that the zeros of $T(z)$ are the zeros of $G(z)$ and the poles of $H(z)$.

Solution

The MATLAB commands for this example are contained in Script 3.4.

MATLAB Script

```
% Script 3.4:  Feedback connection of two blocks
numG = 0.2*[1 -0.85]
a = 0.9*exp(j*pi/4)
denG = conv([1 -a],[1 -conj(a)])     % forward transfer function
G = tf(numG,denG,1)                  % ...with unity sampling period
numH = [1 -0.3]
denH = conv([1 -0.8],[1 -0.1])       % feedback transfer function
H = tf(numH,denH,1)                  % ...with unity sampling interval
T = feedback(G,H)                    % closed-loop system
[zT,pT,kT] = zpkdata(T,'v')          % CL zeros,poles,gain
[resp,dt] = step(T);                 % compute step resp
stem(dt,resp,'filled')               % plot step response with stem option
```

	$G(z)$	$H(z)$	$T(z)$
Zeros	0.85	0.3	$0.1, 0.8, 0.85$
Poles	$0.9\epsilon^{\pm j0.7854}$	$0.1, 0.8$	$0.1503, 0.8124,$ $0.9739\epsilon^{\pm j0.9004}$

TABLE 3.3 *Zeros and poles of the transfer functions in Example 3.4*

The `feedback` command returns the closed-loop transfer function as

$$T(z) = \frac{0.2z^3 - 0.35z^2 + 0.169z - 0.0136}{z^4 - 2.173z^3 + 2.236z^2 - 1.061z + 0.1158}$$

Table 3.3 shows the poles and zeros of $G(z)$, $H(z)$, and $T(z)$, where the complex poles of $G(z)$ and $T(z)$ have been converted to polar form so that their magnitudes are displayed. It is apparent that the zeros of $T(z)$ are the zeros of $G(z)$ and the poles of $H(z)$, and that the poles of $T(z)$ differ from any of the poles of $G(z)$ and $H(z)$. Because the poles of $T(z)$ are all inside the unit circle (exclusive of the unit circle itself), the feedback connection $T(z)$ is asymptotically stable.

WHAT IF? Suppose the forward-path transfer function of the feedback system considered in Example 3.4 contains a gain K such that

$$G(z) = \frac{0.2K(z - 0.85)}{(z - 0.9\epsilon^{j\pi/4})(z - 0.9\epsilon^{-j\pi/4})}$$

Calculate the closed-loop zeros and poles and plot the unit-step response for several values of the gain K in the interval $0.1 \le K \le 2$. Observe that K will affect the closed-loop poles but not the closed-loop zeros. Why is that the case? You should find that the closed-loop system remains stable for gain values up to about $K = 1.3$. ■

□

REINFORCEMENT PROBLEMS

In Problems 3.12 through 3.18, use MATLAB to determine $T(z)$, the transfer function of the closed-loop system obtained by connecting the systems described by $G(z)$ and $H(z)$ with negative feedback, as shown in Figure 3.6. Write $T(z)$ as a ratio of polynomials and make a table showing the zeros and poles of $G(z), H(z)$, and $T(z)$. Determine the stability of the closed-loop system. The sampling period is 1 s.

P3.12 Feedback connection.

$$G(z) = \frac{0.5(3z - 1)}{13z^2 - 5z + 8} \quad \text{and} \quad H(z) = \frac{7z + 4}{(z + 0.5)(4z + 1)}$$

P3.13 Another feedback connection.

$$G(z) = \frac{6(3z+1)}{(z+0.5)(69z^2 - 5z + 5)} \quad \text{and} \quad H(z) = \frac{15z+1}{5z-1}$$

P3.14 One more feedback connection.

$$G(z) = \frac{12z-5}{(10z-0.5)(3z+0.25)} \quad \text{and} \quad H(z) = \frac{6}{3z^2 - 2z - 0.3}$$

P3.15 Unstable open-loop system.

$$G(z) = \frac{1}{(z+1.1)(12z-1)} \quad \text{and} \quad H(z) = \frac{100(z+0.5)}{15z-1}$$

P3.16 Gain variation.

$$G(z) = \frac{K}{z(10z^2 - 2z + 1)}, \quad H(z) = \frac{z-0.8}{2z+1}, \quad \text{and} \quad K = 1$$

Repeat the problem for $K = 2, 5$, and 10. Investigate how the closed-loop system poles vary with the changes in gain K. You should observe that the four closed-loop poles move toward the unit circle as K increases. The pole that moves to the left along the negative real axis passes through $z = -1$ for some $5 < K < 10$, thereby making the closed-loop system unstable. This type of gain-variation investigation, known as *root locus*, is discussed in Appendix C. *Hint:* You can make a vector of the four values of K and create a `for` loop to do the calculations for each of the gains.

P3.17 Gain variation for second-order system.

$$G(z) = \frac{K(3z+0.1)}{(2z+0.1)(z-0.3)}, \quad H(z) = 1, \quad \text{and} \quad K = 0.1$$

Repeat the problem for $K = 0.3, 0.6$, and 1. Investigate how the closed-loop system poles vary as the gain K is increased. Will the feedback system remain stable for all positive values of K? If not, what is the approximate maximum value of K that retains stability?

P3.18 Positive feedback. Repeat Example 3.4 with positive rather than negative feedback. You should find that $T(z)$ has different poles than before but that they are still inside the unit circle. Also $T(z)$ has the same zeros as before.

CONTROLLER TRANSFER FUNCTIONS

In Chapters 8 and 9, we will discuss controllers and techniques for designing a controller to meet a given set of performance specifications. At this point, we present several controller transfer functions so they can be used with the transfer-function models of our "real-world" systems (see Appendix A) to build models of feedback control systems that can be analyzed for stability properties and step response.

We start by defining three basic controllers that can be combined to yield three other, more commonly used controllers. These basic controller types are proportional (P), derivative (D), and integral (I).[1]

The transfer functions of these controllers are expressed in terms of the following parameters:

- proportional gain K_P
- derivative gain K_D, which is often expressed as the derivative time T_D
- integral gain K_I, which is the reciprocal of the reset or integral time T_I
- sampling period T_s

The transfer function of the proportional controller is

$$G_p(z) = K_P \tag{3.3}$$

which is just a constant that equals the proportional gain. We write the transfer function of the derivative controller as

$$G_d(z) = K_P K_D \left(\frac{z-1}{T_s z} \right) \tag{3.4}$$

which has a zero at $z = 1$ and a pole at $z = 0$. Note that this expression includes the proportional gain K_P as a multiplicative factor. This allows us to use the proportional gain as a loop gain that can be adjusted, without affecting the derivative and integral terms, when these basic controllers are combined.

The transfer function of the integral controller can be written as

$$G_i(z) = K_P K_I \left(\frac{T_s z}{z-1} \right) \tag{3.5}$$

which exhibits a pole at $z = 1$ and a zero at $z = 0$, and includes the proportional gain as a factor.

These three basic controllers can be combined to form the proportional-derivative (PD), proportional-integral (PI), and proportional-integral-derivative (PID) controllers. To find the transfer function of the proportional-derivative (PD) controller, we add (3.3) and (3.4) and write the result as a ratio of polynomials in z to obtain

$$G_{\text{pd}}(z) = K_P \left[\frac{(1 + K_D/T_s)z - K_D/T_s}{z} \right] \tag{3.6}$$

[1] Although we are dealing with discrete-time approximations of the continuous-time operations of differentiation and integration, it is customary to use the adjectives *derivative* and *integral* to describe the characteristics of the controller.

The PI controller transfer function is found by adding the two transfer functions (3.3) and (3.5) and forming the result as a polynomial in z, yielding

$$G_{\mathrm{pi}}(z) = K_P \left[\frac{(1 + K_I T_s)z - 1}{z - 1} \right] \tag{3.7}$$

The PID controller transfer function is found by adding the three transfer functions (3.3), (3.4), and (3.5) and forming the result as a polynomial in z, yielding

$$G_{\mathrm{pid}}(z) = K_P \left[\frac{(1 + K_I T_s + K_D/T_s)z^2 - (1 + 2K_D/T_s)z + K_D/T_s}{z(z - 1)} \right] \tag{3.8}$$

In this form, the PID controller can be thought of as a parallel combination of the three controllers described above. Hence, it has poles that are the poles of the above three controllers—namely, a pole at $z = 1$ and a pole at $z = 0$. Its zeros differ from those of the three controllers, however. The PD and PI controllers are special cases of the PID controller; their transfer functions can be obtained from (3.8) by setting $K_I = 0$ or $K_D = 0$, as the case may be.

Useful approximations of these controllers, referred to as *lead, lag,* and *lead-lag compensators*, will be dealt with in Chapter 9.

FEEDBACK SYSTEMS WITH TWO INPUTS

Figure 3.7(a) is a block diagram showing a feedback system with two inputs: (i) the reference input, labeled $R(z)$ and (ii) the disturbance input $D(z)$. Ideally, by applying the controller $G_c(z)$, we would like the plant output $y(k)$ to follow the reference input $r(k)$ and to not be affected by the disturbance input $d(k)$. Although it is possible to construct a 2-input/1-output model of such a system by using the Control System Toolbox `connect` command, we will instead construct two single-input models and use the appropriate one depending on which input is under consideration.

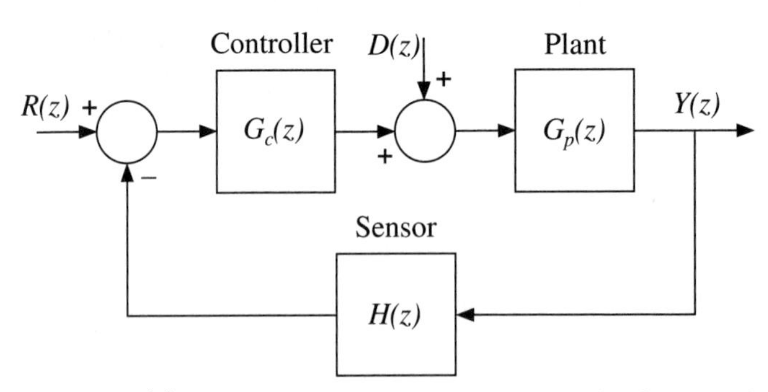

FIGURE 3.7(a) *Block diagram of a feedback system with reference and disturbance inputs: complete system*

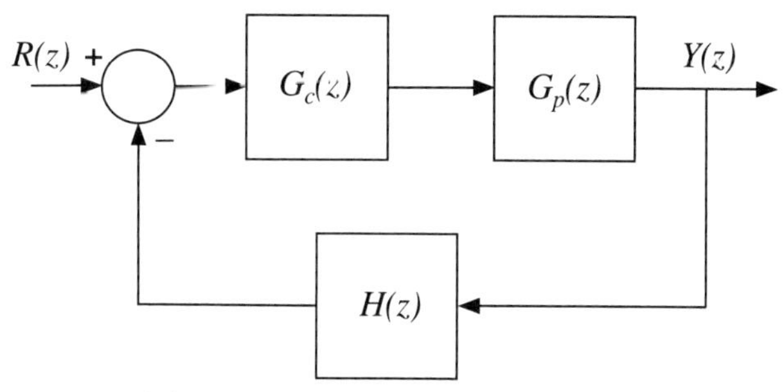

FIGURE 3.7(b) *Block diagram of a feedback system with reference and disturbance inputs: single-input model for reference input only*

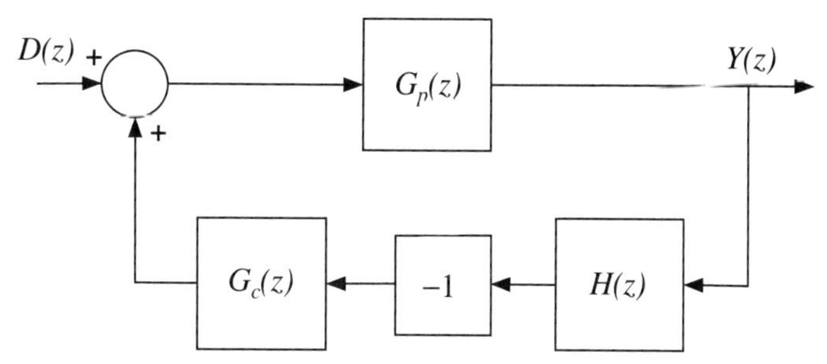

FIGURE 3.7(c) *Block diagram of a feedback system with reference and disturbance inputs: single-input model for disturbance input only*

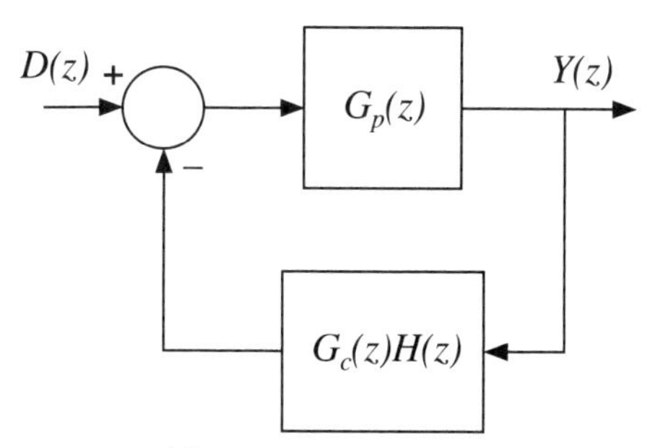

FIGURE 3.7(d) *Block diagram of a feedback system with reference and disturbance inputs: alternate model for disturbance input only*

When we are interested in the response to the reference input, we set the disturbance input to zero and obtain the single-input feedback system shown in Figure 3.7(b). In this case, the forward transfer function is the product of the controller and plant transfer functions $G_p(z)G_c(z)$, and the feedback transfer function is the sensor $H(z)$. With these associations we can use the results of the previous section.

Likewise, when we are interested in the response to the disturbance input, we set $R(z) = 0$ and obtain the single-input block diagram shown in Figure 3.7(c). Now the forward transfer function is $G_p(z)$ and the feedback transfer function is $-G_c(z)H(z)$; there is a plus sign at the arrow where the feedback signal enters the summing junction. Figure 3.7(d) is an equivalent diagram that retains the feature of negative feedback by associating the negative sign in the feedback path with the summing junction.

In Example 3.5, we illustrate the construction of a feedback system having two inputs. The poles of the closed-loop system are unaffected by our choice of input, but the closed-loop zeros are affected. Put another way, the closed-loop system mode functions are the same regardless of the choice of input, but the weightings of these mode functions in the response are different.

☐ EXAMPLE 3.5 *2-Input Feedback System*

Consider a feedback system with both reference and disturbance inputs, as shown in Figure 3.7(a), where the plant and sensor transfer functions are

$$G_p(z) = \frac{0.0031z + 0.003}{z^2 - 1.9z + 0.905} \quad \text{and} \quad H(z) = \frac{0.55}{z - 0.45}$$

and the sampling period is $T_s = 0.04$ s. The controller is an integral controller (3.5) with a proportional gain of $K_P = 10$ and an integral gain of $K_I = 1/20$ s^{-1}. Develop two separate MATLAB models, one model having the reference input $R(z)$ and the other having the disturbance input $D(z)$. Then use these models to determine the closed-loop zeros and poles and to plot their responses to a unit-step function on a single set of axes.

Solution

We begin by using (3.5) to write the controller transfer function as

$$G_c(z) = K_P K_I \left(\frac{T_s z}{z - 1} \right) = \frac{10}{20} \cdot \frac{0.04z}{z - 1} = \frac{0.02z}{z - 1}$$

Next, we implement the three individual blocks as TF objects by entering the row vectors that define their numerator and denominator polynomials or expressions that evaluate them. Then the `feedback` command can be used with the `*` operator to form the series connection in the forward path to obtain the closed-loop model of the reference system shown in Figure 3.7(b). When using the `feedback` command, care must be taken that the correct sign is used for the feedback signal at the summing junction. If a negative sign is desired, it does not have to be shown explicitly. If positive feedback is desired, as in Figure 3.7(c), however, a third argument must be used that has the value `+1`. The `feedback` command is also used to create the disturbance-input model, using the alternate form shown in Figure 3.7(d).

The MATLAB commands in Script 3.5 implement the steps just described for the three transfer functions given in the example statement. In the way of intermediate results, you should find that the closed-loop transfer function for the reference input is

$$T_{\text{ref}}(z) = \frac{(6.2z^3 + 3.21z^2 - 2.7z) \times 10^{-5}}{z^4 - 3.35z^3 + 4.11z^2 - 2.167z + 0.4073}$$

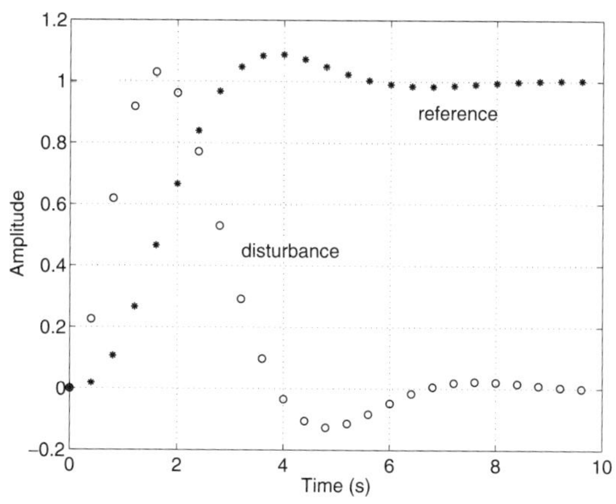

FIGURE 3.8 *Responses of the system in Example 3.5 to steps in the reference and disturbance inputs*

The corresponding transfer function for the disturbance input is

$$T_{\text{dist}}(z) = \frac{(3.1z^3 - 1.495z^2 - 2.955z + 1.35) \times 10^{-3}}{z^4 - 3.35z^3 + 4.11z^2 - 2.167z + 0.4073}$$

Notice that the denominators of the two transfer functions are identical—that is, the transfer functions have the same poles. The numerators are different, however, which means that the step responses are different. This is what we expect, because the system's output $Y(z)$ should follow changes in its reference input $R(z)$ but be unaffected by changes in its disturbance input $D(z)$. Figure 3.8 shows the responses of the system to unit-step functions on the two inputs.

We can compute the steady-state values of the two responses by using the `dcgain` command with the appropriate transfer function as its argument. For the reference transfer function `Tref`, this gives 1.0, which is the steady-state value of the upper curve in Figure 3.8. Using the disturbance transfer function `Tdist` as the argument of the `dcgain` command, we get 0, which is the desired value, because we do not want the plant output to be affected by the disturbance input. The response plot shows that a step-disturbance input causes a transient response, but because of the integral action of the controller, there is no steady-state effect on the output.

Using the `pole` command on either of the closed-loop transfer function LTI objects, `Tref` or `Tdist`, we find the closed-loop poles of the system to be $z = 0.9503, 0.9757\epsilon^{\pm j0.0437}$, and 0.4502, where the complex poles are expressed in polar form.

Recall that the closed-loop zeros are the *zeros* of the *forward* transfer function and the *poles* of the *feedback* transfer function. For the reference transfer function, the closed-loop zeros are $z = 0, -0.9677$, and 0.45, where the first two are the zeros of the forward transfer function (because $G_c(z)$ is in the forward path), and the third is the pole of the feedback transfer function $H(z)$.

For the disturbance transfer function, the zeros are $z = 1, -0.9677$, and 0.45. The zero at $z = 1$ is due to the fact that $G_c(z)$ is in the feedback path for the disturbance input. The zero at $z = 0.45$ is due to the fact that the sensor transfer function $H(z)$ is in the feedback path in this configuration as well, and it has a pole there. The zero at $z = -0.9677$ is due to $G_p(z)$.

MATLAB Script

```
% Script 3.5:  Feedback system with reference and disturbance inputs
%-- define:  sampling period, controller gain, integral time constant
Ts = 0.04, Kp = 10, Ki = 1/20
Gp = tf([0.0031 0.003],[1 -1.9 0.905],Ts)   % plant as TF object
H  = tf(0.55,[1 -0.45],Ts)                  % sensor as TF object
Gc = tf([Kp*Ki*Ts 0],[1 -1],Ts)             % integral controller
%--------- CL transfer function for reference input ------
Tref = feedback(Gp*Gc,H)
[zTref,pTref,kTref] = zpkdata(Tref,'v')     % zeros, poles, gain
Tref_gainDC = dcgain(Tref)                  % reference DC gain
dtime = [0:Ts:10];              % discrete time vector for 10 seconds
yref = step(Tref,dtime);        % response to step in reference input
%
%----------- repeat for disturbance input -----------
Tdist = feedback(Gp,Gc*H)
[zTdist,pTdist,kTdist] = zpkdata(Tdist,'v') % zeros, poles, gain
Tdist_gainDC = dcgain(Tdist)                % disturbance DC gain
ydist = step(Tdist,dtime);     % response to step in disturbance input
%----- plot step responses at every 10th point ---------------
kk = [1:10:10/Ts];
plot(dtime(kk),yref(kk),'*')            % plot reference step response
grid,hold on
plot(dtime(kk),ydist(kk),'o')           % plot disturbance step response
hold off
text(6.5,0.95,'reference')              % add labels to plots
text(3.2,0.50,'disturbance')
```

WHAT IF? Replace the gain $K_P = 10$ in the numerator of $G_c(z)$ in Example 3.5 with the variable gain K_P, whose value you can specify with the `input` command, as in `Kp = input('enter controller gain....')`. Then compute and plot the responses to both reference and disturbance step inputs for several values of `Kp`. You should find that as the gain increases (i) the overshoot increases and the response becomes more oscillatory, (ii) the initial response is faster, and (iii) the steady-state errors are reduced. By trial and error, estimate the maximum value of K_P for a stable closed-loop system. Also try reducing the gain. How are the overshoot and response time affected? ■

□

FEEDBACK CONNECTIONS OF MULTIPLE-INPUT, MULTIPLE-OUTPUT SYSTEMS

In the previous section, we demonstrated the use of the MATLAB `feedback` function to make connections for single-input, single-output systems. For systems with more than one input or output, the `feedback` command allows selective closure of feedback loops. Consider the two feedback connections shown in Figure 3.9 for a single-input, two-output plant $G(z)$, in which only the $y_2(k)$ signal is fed back. Figure 3.9(a) shows the controller $G_c(z)$ in the feedback path, which is commonly used for stabilization, whereas Figure 3.9(b) shows $G_c(z)$ in the forward path, which is commonly used for regulation. The `feedback` function has the flexibility to allow for both of these connections. This is illustrated in Example 3.6, which is related to the system used in Example 2.12.

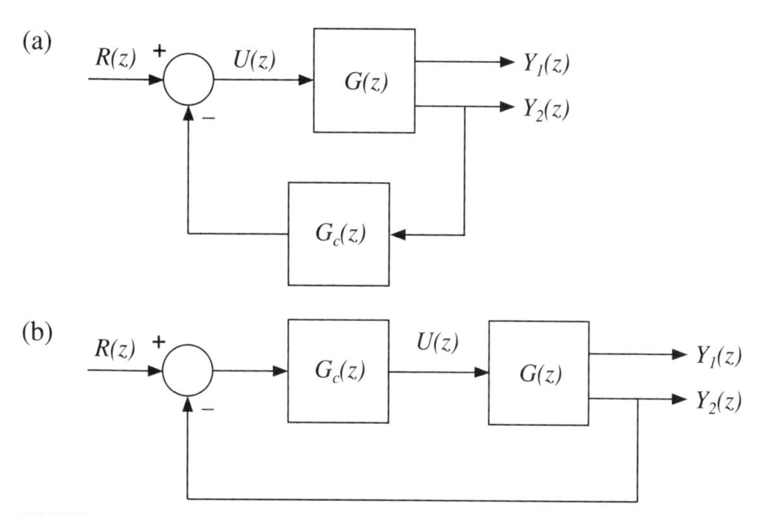

FIGURE 3.9 *Feedback connections with (a) controller in feedback path; (b) controller in forward path*

☐ EXAMPLE 3.6 *Feedback Connection Using a Selective Output*

Find the closed-loop system transfer functions shown in Figure 3.9 for

$$G(z) = \begin{bmatrix} G_1(z) \\ G_2(z) \end{bmatrix} = \begin{bmatrix} \dfrac{0.0134}{(z-0.9394)(z-0.7788)} \\ \dfrac{0.2212}{z-0.7788} \end{bmatrix}$$

and the proportional controller $G_c(z) = 3$. For each case, the closed-loop transfer function should be of dimension 2×1, implying one input and two outputs. The sampling period is 0.05 s.

Solution

In the first part of Script 3.6, the system and controller are entered as `G` and `Gc`. For part a, in the command `Gcla = feedback(G,Gc,1,2)`, the first input argument `G` is the LTI object in the forward path, the second input argument `Gc` is the LTI object in the feedback path, and the third and fourth input arguments designate the indices of the input and output used for the feedback loop. For part b, the command `Gclb = feedback(G*Gc,1,1,2)` is similar to the one used in part a, except that both `G` and `Gc` are in the forward path and the feedback path is unity. It is important that the output of `Gc` connects to the input of `G`, as denoted by `G*Gc`. Because the feedback function using transfer functions may introduce redundant closed-loop poles and zeros that will cancel one another, we apply the `minreal` function to eliminate them. The resulting closed-loop transfer function for part a is

$$G_a(z) = \begin{bmatrix} \dfrac{0.0134}{z^2 - 1.055z + 0.1082} \\ \dfrac{0.2212}{z - 0.1152} \end{bmatrix}$$

and for part b is

$$G_b(z) = \begin{bmatrix} \dfrac{0.0402}{z^2 - 1.055z + 0.1082} \\ \dfrac{0.6636}{z - 0.1152} \end{bmatrix}$$

Although the two feedback configurations are different, the transfer functions have the same poles, albeit different gains. The step responses of the transfer functions are shown in Figure 3.10. Compared to the open-loop step responses in Figure 2.12, the y_2 response has been made much faster, and the y_1 response has also been made slightly faster in both cases. Because there is no integral action in the feedback loop, both outputs are different from unity.

MATLAB Script

```
% Script 3.6:  Feedback connection using a selective input
Ts = 0.05                             % sampling time
% enter 1-input/2-output system
G1 = tf(0.01340,conv([1 -0.9394],[1 -0.7788]),Ts)
G2 = tf(0.2212,[1 -0.7788],Ts)
G  = [G1; G2]                         % adjoin system
Gc = 3
% ---- part a:  controller in feedback path ------
% Gc connects from output 2 to input 1
Gcla = feedback(G,Gc,1,2)
Gcla = minreal(Gcla)                  % eliminate redundant poles and zeros
dtime = 0:Ts:4;                       % time vector
ya = step(Gcla,dtime);                % step response
plot(dtime,ya(:,1),'o',dtime,ya(:,2),'*') % plot y1 & y2
% ---- part b:  controller in forward path ------
% Gc connects from output 2 to input 1
Gclb = feedback(G*Gc,1,1,2)
Gclb = minreal(Gclb)                  % eliminate redundant poles and zeros
yb = step(Gclb,dtime);                % step response
plot(dtime,yb(:,1),'o',dtime,yb(:,2),'*') % plot y1 & y2
```

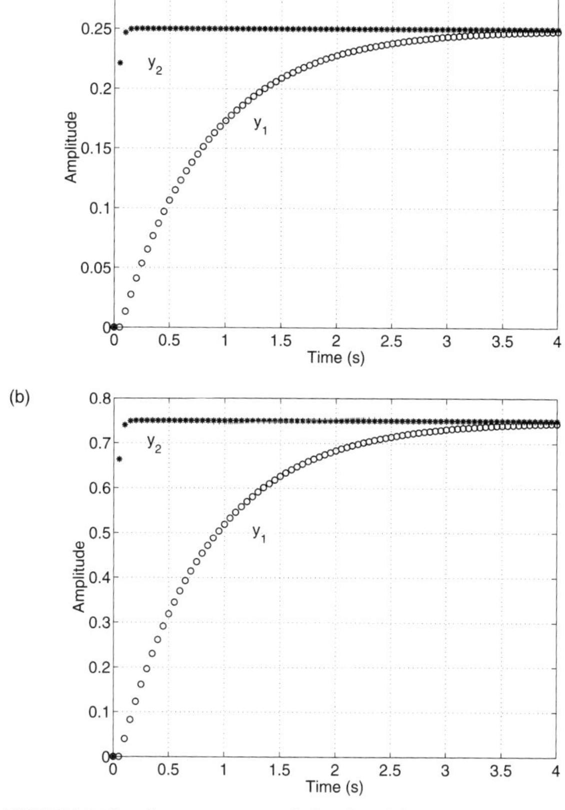

FIGURE 3.10 *Step responses of the closed-loop systems in Example 3.6: (a) G_c in the feedback path; (b) G_c in the forward path*

□

EXPLORATORY PROBLEMS

EP3.1 Interactive analysis with MATLAB**.** The MATLAB file `done_blk.m` can be used to perform a variety of analytical tasks on any single-block system whose transfer function exists in the workspace as the row vectors `num` and `den`. You can create your own one-block models as described in the statement of Exploratory Problem EP2.1 in Chapter 2.

Alternatively, you can select the **Activate keyboard** entry on the model-building menu. At this point you can (i) enter your own model, (ii) create interconnections of models, or (iii) create a TF model of one of the real-world systems described in Appendix A by running the appropriate M-file and selecting the desired transfer function on the menu. At the conclusion of the M-file, the transfer function of the plant or process will be in the workspace. If you rename the vectors that contain the coefficients of the numerator and denominator polynomials of the closed-loop system as **num** and **den**, you can use **done_blk.m** to perform extensive analysis without having to enter more MATLAB commands.

To summarize, you can create models of almost any single-input/single-output system you wish and then use **done_blk.m** to do the analysis, provided only that you use **num** and **den** as the names for the numerator and denominator coefficient vectors.

COMPREHENSIVE PROBLEMS

CP3.1 Ball and beam system. Consider the design of two controllers, $K_m(z)$ and $K_b(z)$, shown in Figure A.3 in Appendix A for the ball and beam system. First, run the **dbbeam.m** file to obtain a discrete-time model of the system sampled at $T_s = 0.02$ s. Then construct a proportional controller for the servomotor loop by setting $K_m(z) = K_{P1}$. Select the gain K_{P1} to be in the range of 0.5 to 5 and examine the closed-loop response of the wheel angle θ up to 2 s for a step command of 0.2 rad. Because this is a one-input, two-output system, closing the $K_m(z)$ loop requires the MATLAB command `G_Kmcl = feedback(Gztf*Km,1,1,2)`. The controller `Km` in the forward loop is connected to the input of the 2×1 ball and beam model (first argument), the feedback loop is unity gain (second argument), and the feedback connection is from the second output (fourth argument) to the first input (third argument). (You can enter **help feedback** in the MATLAB command window to learn more about options for the **feedback** function.) Select a value of $K_m(z) = K_{P1}$ such that a fast response with no overshoot is obtained.

With the $K_m(z)$ loop closed—that is, using the model `G_Kmcl(1,1)` —set $K_b(z)$ to be a proportional controller with the gain K_{P2} from 0.5 to 5 and examine the closed-loop system response to a position command of 0.1 m. You will notice that the system becomes unstable for high values of K_{P2}. To improve the system response, replace $K_b(z)$ with a PD controller described by (3.6) using $K_{P2} = 1$ and $K_D = 1$, and repeat the closed-loop system response. Discuss the improvements in the response of the PD controller over the P controller.

CP3.2 Inverted pendulum with angle feedback. Construct a feedback system for the discrete-time inverted pendulum model by feeding back the stick angle θ with a proportional-derivative (PD) controller as given by (3.6). The transfer function of the plant from the input to the pendulum angle with a sampling period of $T_s = 0.01$ s can be obtained by running `dstick.m`. Start by drawing a pole-zero plot for the open-loop system. This should show that the plant model has a pole to the right of the zero at $z = 1.0$. Using a derivative gain of $K_D = 0.2$ and a proportional gain of $K_P = 1.5$, compute the closed-loop poles and plot the step response to a change in desired pendulum angle. You should find that the closed-loop system is unstable. Furthermore, you will be unable to find a combination of values for K_D and K_P that yields a stable system. In Chapter 8, we will use root-locus plots to show why this is the case.

CP3.3 Electric power generation system. Consider the voltage control of the electric power system shown in Figure A.8 in Appendix A, using a proportional-integral (PI) controller with its transfer function $K_V(z)$ as given by (3.7). Set the integral gain at $K_I = 1$, vary the controller gain K_P in the range of 10 to 50, and obtain the transfer function of the closed-loop system. Determine the poles, zeros, and gain of the closed-loop system, and plot its response to a step of amplitude 0.05 in the reference voltage input V_{ref}. You should find that for the closed-loop system, the lightly damped electromechanical mode becomes unstable at a gain of about $K_P = 42$. The instability can be observed as growing oscillations about the steady-state value of the step response.

CP3.4 Hydroturbine system. Construct a feedback system for controlling the mechanical power produced by the hydroturbine, as illustrated in Figure A.10 in Appendix A, where $G(z)$ denotes the discrete-time transfer function of the series combination of the gate actuator and the hydroturbine unit. In each case, obtain the transfer function of the closed-loop system with the specified controller and parameter values; determine the poles, zeros, and DC gain of the closed-loop system; and plot its response to a unit step in the desired power. The sampling period is $T_s = 0.1$ s.

a. Use a proportional controller with $K_P = 0.5$. You should find that the closed-loop system is stable, but the response reaches a steady-state value of only 0.333, whereas a steady-state response of 1.0 is desired. Also show that the system becomes marginally stable when the gain is increased to 0.735. Observe that the step response starts in the negative direction, just as we saw in Figure 2.10 of Example 2.11. This unusual behavior is due to the presence of a zero on the positive real axis, *outside* the unit circle.

b. Use a proportional-derivative (PD) controller, as defined in (3.6), with $K_P = 0.5$ and $K_D = 0.3$. You should find that for a unit-step input, the steady-state response is still 0.333, but the response settles faster than it does for the proportional controller. Experiment with the gains to see if you can improve the response obtained with the suggested gains.

c. Try a proportional-integral (PI) controller, as defined in (3.7), with $K_P = 0.5$ and $K_I = 0.4$. You should find that the steady-state response is now unity, although the response is somewhat slower than with the PD controller, and the initial response is still negative. In later chapters, we will consider ways to improve these responses.

SUMMARY

In this chapter we have shown how to combine transfer-function models of two or more subsystems to construct more complicated and versatile system models. We have considered series, parallel, and feedback combinations of models that are described by transfer functions in the TF form, namely as a ratio of polynomials. Once the transfer function of the combined system is available, it is possible to perform analytical calculations such as finding the zeros, poles, and gain or establishing the stability of the system. It is also possible to obtain plots of the response to step functions, impulses, or arbitrary inputs. In Chapter 4, we will show how MATLAB can be used to construct models of more complicated systems in state-space form.

The Control System Toolbox has a command named `connect` for building models having a more general structure than those in the examples we have considered here, but we have not illustrated it.

ANSWERS

P3.1 Zeros of $T(z)$ are $z = -0.2500$ and $0.5\epsilon^{\pm j0.7855}$; poles of $T(z)$ are $z = 0.7\epsilon^{\pm j1.0472}$ and $0.8\epsilon^{\pm j1.2569}$; gain of $T(z)$ is 2. Stable.

P3.2 Zero of $T(z)$ is $z = 0.4$; poles of $T(z)$ are $z = -0.1, 0.5385\epsilon^{\pm j1.1903}$, and 0.8; gain of $T(z)$ is 12. Stable.

P3.3 Zeros of $T(z)$ are $z = -0.2$ and -0.7; poles of $T(z)$ are $z = -0.8, 0.9\epsilon^{\pm j0.6283}$, 0.8, and 0.9; gain of $T(z)$ is 0.02. Stable.

P3.4 Zeros of $T(z)$ are $z = -1.5, -0.3333$, and -0.1; poles of $T(z)$ are $z = -0.5, 0.6325\epsilon^{\pm j1.8925}$ and $0.8944\epsilon^{\pm j1.6828}$; gain of $T(z)$ is 12. Stable.

P3.5 After cancellation, zero of $T(z)$ is $z = -0.1667$; poles of $T(z)$ are $z = 0.3536\epsilon^{\pm j1.2094}$ and $0.3536\epsilon^{\pm j2.3562}$; gain of $T(z)$ is 6. Stable.

P3.6 After cancellation, zeros of $T(z)$ are $z = -0.4$ and 0.6; poles of $T(z)$ are $z = -0.1, 0.125\epsilon^{\pm j1.5708}$ and 0.5; gain of $T(z)$ is 1.5625. Unstable, due to the cancellation of the unstable pole $z = -1.2$ of $G_1(z)$.

P3.7 $T(z) = (15z^3 + 25.5z^2 + 5.2z + 3.3)/(30z^3 + 7z^2 + 10z - 2)$

P3.8 $T(z) = (6z^4 + 23.2z^3 + 10.4z^2 + 0.6z - 4)/(8z^4 + 7.6z^3 + 9.2z^2 + 3.8z + 2)$

P3.9 After a stable pole-zero cancellation at $z = 0.2$, $T(z)$ has zeros at $z = 0.3162\epsilon^{\pm j1.7296}$ and poles at $z = -0.3, -0.1$, and 0.2. Stable.

P3.10 Zeros are $z = -0.6667, -0.2495, 0.6970$, and $0.4753\epsilon^{\pm j2.3965}$; poles are $z = -0.8660, -0.2, 0.5, 0.8, 0.8660$, and $0.5\epsilon^{\pm j2.2459}$; gain is 5.25. Stable.

P3.11 Zeros are $z = 0.2342\epsilon^{\pm j2.6096}$ and $1.6876\epsilon^{j1.7511}$; poles are $z = -0.3, 0.0625, 0.9\epsilon^{\pm j1.4033}$, and $1.0\epsilon^{\pm j2.0944}$; gain is 2.8. Marginally stable.

P3.12 $T(z) = (6z^3 + 2.5z^2 - 0.75z - 0.25)/(52z^4 + 19z^3 + 34z^2 + 24z + 2)$

P3.13 $T(z) = (90z^2 + 12z - 6)/(345z^4 + 78.5z^3 + 253z^2 + 118z + 3.5)$

P3.14 $T(z) = (36z^3 - 39z^2 + 6.4z + 1.5)/(90z^4 - 57z^3 - 11.38z^2 + 71.95z - 29.96)$

P3.15 Zero is $z = 0.0667$; poles are $z = -0.8698$ and $0.5713\epsilon^{\pm j1.6410}$; gain is 0.0833. Stable.

P3.16 For $K = 5$, the zero is $z = -0.5$; poles are $z = -0.9167, 0.4460$, and $0.6994\epsilon^{\pm j1.4485}$; gain is 0.5. Stable for $K = 1, 2$, and 5; unstable for $K = 10$.

P3.17 For $K = 0.6$, the zero is $z = -0.0333$; poles are $z = -0.0240$ and -0.6260; gain is 0.9. Stable for $K = 0.1, 0.3$, and 0.6. Unstable for $K = 1$.

P3.18 $T(z) = (0.2z^3 - 0.35z^2 + 0.169z - 0.0136)/(z^4 - 2.173z^3 + 1.836z^2 - 0.6008z + 0.0138)$. Poles are $z = 0.0248, 0.7752$, and $0.8474\epsilon^{\pm j0.6266}$. Stable.

4 State-Space Models

PREVIEW

The use of transfer functions to represent linear, time-invariant systems is suitable when the model is given in the input-output form and when the model order is low. Models of real systems are usually derived from physical laws in terms of state variables that correspond to identifiable quantities such as stored energies, however. Such models are put in the state-space form and can be compactly represented in matrix notation. One advantage of the state-space form is that the effects of nonzero initial conditions on the system response can be readily investigated. In addition, for high-order systems, the state-space form is preferable because it is less susceptible to numerical ill-conditioning than the transfer-function form. There are Control System Toolbox commands available to accept the model of a system represented in any of the three forms: transfer-function, zero-pole-gain, or state-space. In this chapter we apply the MATLAB commands discussed in Chapters 2 and 3 to build and analyze state-space models.

MODEL BUILDING, CONVERSIONS, AND INTERCONNECTIONS

For a linear, time-invariant discrete-time system with n states, m inputs, and p outputs, the state-space model in matrix notation is

$$\mathbf{x}(k+1) = \mathbf{A}\mathbf{x}(k) + \mathbf{B}\mathbf{u}(k), \quad \mathbf{y}(k) = \mathbf{C}\mathbf{x}(k) + \mathbf{D}\mathbf{u}(k) \tag{4.1}$$

where the state vector $\mathbf{x}(k)$ is of dimension n, the input vector $\mathbf{u}(k)$ of dimension m, and the output vector $\mathbf{y}(k)$ of dimension p. The state matrix $\mathbf{A}$ is of dimension $n \times n$, the input matrix $\mathbf{B}$ is $n \times m$, the output matrix $\mathbf{C}$ is $p \times n$, and the feedforward matrix $\mathbf{D}$ is $p \times m$. For a time-invariant system, $\mathbf{A}$, $\mathbf{B}$, $\mathbf{C}$, and $\mathbf{D}$ are constant matrices.

Equation (4.1) is written in terms of the time variable k, which takes on non-negative integer values, namely $0, 1, 2, 3, \ldots$. Should it be important to know the values of the variables at specific points in time, we use the relationship $t = kT_s$, where T_s is the sampling period, to make the conversion from discrete time to "real" time.

Because (4.1) is the state-space representation, we will frequently refer to it as the SS form. The development of a state-space model from a physical system can be found in many control systems engineering textbooks. Once the matrices $\mathbf{A}$, $\mathbf{B}$, $\mathbf{C}$, and $\mathbf{D}$ are known, the state-space model can be readily entered in MATLAB by defining the four matrices and using the `ss` command. For example, if we have defined the matrices $\mathbf{A}$, $\mathbf{B}$, $\mathbf{C}$, and $\mathbf{D}$ as the MATLAB variables `A, B, C,` and `D`, the model can be created as a state-space (SS) object by entering `Gss = ss(A,B,C,D,Ts)`, where `Ts` is either set to the sampling period or to `-1` to denote an unspecified sampling period.

The state-space model (4.1) has the matrix transfer function $\mathbf{G}(z)$ which obeys the relation

$$\mathbf{G}(z) = \mathbf{C}(z\mathbf{I} - \mathbf{A})^{-1}\mathbf{B} + \mathbf{D}$$

where $\mathbf{I}$ is the identity matrix of the same dimension as $\mathbf{A}$. If a system model `G1` exists in TF form, a state-space equivalent can be obtained by entering `G1ss = ss(G1)`. Keep in mind that there is not a unique state-space representation for a given transfer function. In fact, the number of state variables—that is, the number of rows or columns of $\mathbf{A}$—may vary from one representation to another. The state-space representation obtained using the `ss` command is guaranteed to be minimal only if the system is either single-input or single-output.

In similar fashion, if a system model `G2` exists in ZPK form, a state-space equivalent can be obtained by entering `G2ss = ss(G2)`. We discuss equivalent state-space models in a later section on state-variable transformations.

The Control System Toolbox provides the `ssdata` function for extracting properties of a SS object. For example, we would enter `[A,B,C,D,Ts] = ssdata(S1)` to obtain the four matrices and the sampling time describing the state-space system `S1`. We can also use the `ssdata` command on TF or ZPK objects. In these cases, the TF or ZPK object is first converted implicitly to SS form.

The commands for interconnecting models introduced in Chapter 3 for TF and ZPK systems can be used in exactly the same manner with systems expressed as SS objects. Series connections are made with the `*` operator; particular attention must be paid to the ordering of the systems. Parallel connections are made with the `+` operator; here, attention must be paid to the number of inputs and outputs when dealing with multi-input/multi-output (MIMO) systems. The `feedback` command allows the interconnection of subsystems in state-space or other forms. Again, care must be taken when working with MIMO systems to ensure that the intended input and output pairings are being made.

One additional consideration is that the user must be aware of the precedence rules when there is a mixture of system object types. Because the state-space representation is the best of the three in terms of numerical reliability, the resulting object will be SS if at least one of the subsystems being connected is of type SS. If none of the subsystems is SS, but at least one is of type ZPK, the resulting object will be ZPK. The overall connection will be of type TF only if all of its components are TF objects.

The following two examples illustrate how to enter a SS form of a system, to convert it to the TF form, and to perform interconnections using the SS form.

☐ EXAMPLE 4.1 *Building a State-Space Model*

Enter in MATLAB the matrices

$$\mathbf{A} = \begin{bmatrix} 0.98 & 0.095 & 0.003 \\ -0.38 & 0.89 & 0.04 \\ 0 & 0 & 0.14 \end{bmatrix}, \quad \mathbf{B} = \begin{bmatrix} 0.002 \\ 0.05 \\ 0.9 \end{bmatrix}$$

$$\mathbf{C} = \begin{bmatrix} 1 & 0 & 0 \end{bmatrix}, \quad \mathbf{D} = \begin{bmatrix} 0 \end{bmatrix}$$

and specify the sampling period as $T_s = 0.1$ s. Next, create the model as a SS object. Finally, convert the model to TF form and then back to SS form, using a different name. Then convert the original SS form to ZPK form and back to SS form.

Solution

MATLAB allows the entry of a matrix on a single line, with the semicolon denoting the end of a row. This feature is used to enter the matrices `A`, `B`, `C`, and `D`, as shown in Script 4.1. The SS object named `Gss` is created by the command `Gss = ss(A,B,C,D,Ts)`, which displays the four matrices, with the columns and rows labeled according to the particular state, input, or output variable involved, using the generic notation `x1, x2, x3` for the three states, `u1` for the single input, and `y1` for the single output.

The command `Gtf = tf(Gss)` creates the system as a TF object and causes the transfer function to be displayed as the ratio of its numerator and denominator polynomials—namely,

$$G(z) = \frac{(2.0z^2 + 5.39z + 0.6012) \times 10^{-3}}{z^3 - 2.01z^2 + 1.17z - 0.1272}$$

The command `Gss1 = ss(Gtf)` converts the TF object back to SS form, with the name `Gss1`, to make it distinguishable from the original SS object `Gss`. The resulting matrices are

$$\mathbf{A}_1 = \begin{bmatrix} 2.01 & -0.5850 & 0.2543 \\ 2 & 0 & 0 \\ 0 & 0.25 & 0 \end{bmatrix}, \quad \mathbf{B}_1 = \begin{bmatrix} 0.0625 \\ 0 \\ 0 \end{bmatrix}$$

$$\mathbf{C}_1 = \begin{bmatrix} 0.032 & 0.0431 & 0.0192 \end{bmatrix}, \quad \mathbf{D}_1 = \begin{bmatrix} 0 \end{bmatrix}$$

These clearly differ from the original set used to create `Gss`.

The form of $\mathbf{A}_1$ and $\mathbf{B}_1$ is a variant of the more familiar *controller form.* In $\mathbf{A}_1$, the lower diagonal entries are not all ones, as is the case for the controller form. Although these two representations are not the same, commands discussed in Example 4.3 can be used to show they have the same zeros, poles, and DC gain. This means that the two systems are equivalent from an input-output point of view.

Next, we convert the original SS object `Gss` to ZPK form with the command `Gzpk = zpk(Gss)`. The system's transfer function is displayed in zero-pole-gain form as

$$G(z) = \frac{0.002(z + 2.578)(z + 0.1166)}{(z - 0.14)(z^2 - 1.87z + 0.9083)}$$

Finally, we convert this ZPK form back to SS form by entering `Gss2 = ss(Gzpk)`. The resulting state-space matrices are

$$\mathbf{A}_2 = \begin{bmatrix} 0.9350 & 0.5000 & 0.1179 \\ -0.0682 & 0.9350 & 0.1892 \\ 0 & 0 & 0.1400 \end{bmatrix}, \quad \mathbf{B}_2 = \begin{bmatrix} 0 \\ 0 \\ 0.3578 \end{bmatrix}$$

$$\mathbf{C}_2 = \begin{bmatrix} 0.2164 & 0 & 0.0056 \end{bmatrix}, \quad \mathbf{D}_2 = \begin{bmatrix} 0 \end{bmatrix}$$

Note that $\mathbf{A}_2$ is in the upper block-triangular form, in which its diagonal blocks are of dimension either 1×1 or 2×2. The 2×2 blocks are for complex poles. Clearly, these matrices are quite different from the two other sets. They do represent the same transfer function from input to output, however. As such, they provide a valid state-space representation of the original system.

MATLAB Script

```
% Script 4.1:  Entering a state-space model
Ts = 0.1                                  % sampling period
A = [0.98  0.095  0.003; -0.38  0.89  0.04; 0  0  0.14]
B = [0.002; 0.05; 0.9]
C = [1 0 0]
D = 0
Gss = ss(A,B,C,D,Ts)                      % Set up model in SS form
Gtf = tf(Gss)                             % convert to TF form
Gss1 = ss(Gtf)                            % convert back to SS form
[A1,B1,C1,D1] = ssdata(Gss1)              % extract state-space matrices
Gzpk = zpk(Gss)                           % convert to ZPK form
Gss2 = ss(Gzpk)                           % convert back to SS form
[A2,B2,C2,D2] = ssdata(Gss2)              % extract state-space matrices
```

□

REINFORCEMENT PROBLEMS

For each of the following problems, enter the state-space matrices of the system into MATLAB and create the model as an SS object. Then convert the model to TF form and then back to SS form, and compare the **A**, **B**, **C**, and **D** matrices of the two forms. Repeat the process using the ZPK form as the intermediate one.

P4.1 Third-order system.

$$\mathbf{A} = \begin{bmatrix} 0.85 & -0.56 & 0.22 \\ 0.20 & 0.11 & 1 \\ -0.25 & 0.50 & 0 \end{bmatrix}, \quad \mathbf{B} = \begin{bmatrix} 1 \\ 0 \\ 0 \end{bmatrix}$$

$$\mathbf{C} = \begin{bmatrix} 0.58 & -0.30 & 0.16 \end{bmatrix}, \quad \mathbf{D} = \begin{bmatrix} 0 \end{bmatrix}$$

The sampling period is $T_s = 0.1$ s.

P4.2 Fourth-order system with $\mathbf{D} \neq 0$.

$$\mathbf{A} = \begin{bmatrix} -0.819 & 0.148 & 0.0277 & 0.00398 \\ 0 & 0.670 & 0.255 & 0.0535 \\ 0 & 0 & 0.606 & 0.238 \\ 0 & 0 & 0 & 0.368 \end{bmatrix}, \quad \mathbf{B} = \begin{bmatrix} 0.00111 \\ 0.021 \\ 0.155 \\ 0.632 \end{bmatrix}$$

$$\mathbf{C} = \begin{bmatrix} 1 & 0 & 1 & 0 \end{bmatrix}, \quad \mathbf{D} = \begin{bmatrix} 0.5 \end{bmatrix}$$

The sampling period is $T_s = 0.5$ s.

☐ EXAMPLE 4.2 *Series, Parallel, and Feedback Connections*

Two systems with transfer functions $G_1(z)$ and $G_2(z)$ and a sampling period of unity are defined by the state-space models

$$\mathbf{A}_1 = \begin{bmatrix} -0.8 & -0.97 \\ 1 & 0 \end{bmatrix}, \quad \mathbf{B}_1 = \begin{bmatrix} 0.45 \\ 0 \end{bmatrix}$$

$$\mathbf{C}_1 = \begin{bmatrix} 0.44 & -0.2 \end{bmatrix}, \quad \mathbf{D}_1 = \begin{bmatrix} 0 \end{bmatrix}$$

$$\mathbf{A}_2 = \begin{bmatrix} 0.95 & 1 \\ 0 & 0.8 \end{bmatrix}, \quad \mathbf{B}_2 = \begin{bmatrix} 0 \\ 1 \end{bmatrix}$$

$$\mathbf{C}_2 = \begin{bmatrix} 0.3 & 0 \end{bmatrix}, \quad \mathbf{D}_2 = \begin{bmatrix} 0.2 \end{bmatrix}$$

Find state-space models for the following interconnections:

a. series: $G_s(z) = G_2(z)G_1(z)$

b. parallel: $G_p(z) = G_1(z) + G_2(z)$

c. feedback: $G_f(z) = \dfrac{G_1(z)}{1 + G_1(z)G_2(z)}$

Solution

The MATLAB commands in Script 4.2 will compute the series, parallel, and feedback connections of $G_1(z)$ and $G_2(z)$, expressed in the SS form.

MATLAB Script

```
% Script 4.2:  Interconnections in state-space form
Ts = 1                          % unity sampling period
A1 = [-0.8 -0.97; 1 0]          % state-space matrices for G1
B1 = [0.45; 0]
C1 = [0.44 -0.2]
D1 = 0
G1 = ss(A1,B1,C1,D1,Ts)         % build G1 as SS object
A2 = [0.95 1; 0 0.8]            % state-space matrices for G2
B2 = [0; 1]
C2 = [0.3 0]
D2 = 0.2
G2 = ss(A2,B2,C2,D2,Ts)         % build G2(z) as SS object
Gs = G2*G1                      % series connection Gs(z)
Gp = G1+G2                      % parallel connection Gp(z)
Gf = feedback(G1,G2)            % feedback connection Gf(z)
```

The series connection results in the matrices

$$\mathbf{A}_s = \begin{bmatrix} 0.95 & 1 & 0 & 0 \\ 0 & 0.8 & 0.44 & -0.2 \\ 0 & 0 & -0.8 & -0.97 \\ 0 & 0 & 1 & 0 \end{bmatrix}, \quad \mathbf{B}_s = \begin{bmatrix} 0 \\ 0 \\ 0.45 \\ 0 \end{bmatrix}$$

$$\mathbf{C}_s = \begin{bmatrix} 0.3 & 0 & 0.088 & -0.04 \end{bmatrix}, \quad \mathbf{D}_s = \begin{bmatrix} 0 \end{bmatrix}$$

The parallel connection results in the matrices

$$\mathbf{A}_p = \begin{bmatrix} -0.8 & -0.97 & 0 & 0 \\ 1 & 0 & 0 & 0 \\ 0 & 0 & 0.95 & 1 \\ 0 & 0 & 0 & 0.8 \end{bmatrix}, \quad \mathbf{B}_p = \begin{bmatrix} 0.45 \\ 0 \\ 0 \\ 1 \end{bmatrix}$$

$$\mathbf{C}_p = \begin{bmatrix} 0.44 & -0.2 & 0.3 & 0 \end{bmatrix}, \quad \mathbf{D}_p = \begin{bmatrix} 0.2 \end{bmatrix}$$

Note the special structures of these state matrices. For example, the main 2×2 diagonal blocks of $\mathbf{A}_s$ and $\mathbf{A}_p$ are $\mathbf{A}_1$ and $\mathbf{A}_2$.

The feedback connection results in the closed-loop system matrices

$$\mathbf{A}_f = \begin{bmatrix} -0.8396 & -0.952 & -0.135 & 0 \\ 1 & 0 & 0 & 0 \\ 0 & 0 & 0.95 & 1 \\ 0.44 & -0.2 & 0 & 0.8 \end{bmatrix}, \quad \mathbf{B}_f = \begin{bmatrix} 0.45 \\ 0 \\ 0 \\ 0 \end{bmatrix}$$

$$\mathbf{C}_f = \begin{bmatrix} 0.44 & -0.2 & 0 & 0 \end{bmatrix}, \quad \mathbf{D}_f = \begin{bmatrix} 0 \end{bmatrix}$$

Because of the presence of feedback, the structure of $\mathbf{A}_f$ is more complicated than that of its counterparts $\mathbf{A}_s$ and $\mathbf{A}_p$.

□

REINFORCEMENT PROBLEMS

In Problems 4.3 and 4.4, find the SS forms of the series, parallel, and feedback connections for the systems $G_1(z)$ and $G_2(z)$. The sampling period is 0.2 s. Verify that all three connections are stable.

P4.3 $G_1(z)$ and $G_2(z)$ given in SS form.

$$\mathbf{A}_1 = \begin{bmatrix} 0.98 & 0.18 & 0.014 \\ -0.22 & 0.71 & 0.11 \\ -1.76 & -2.42 & 0.16 \end{bmatrix}, \quad \mathbf{B}_1 = \begin{bmatrix} 0.001 \\ 0.014 \\ 0.11 \end{bmatrix}$$

$$\mathbf{C}_1 = \begin{bmatrix} 4 & 2 & 0 \end{bmatrix}, \quad \mathbf{D}_1 = \begin{bmatrix} 0 \end{bmatrix}$$

$$\mathbf{A}_2 = \begin{bmatrix} 0.2 \end{bmatrix}, \quad \mathbf{B}_2 = \begin{bmatrix} 0.1 \end{bmatrix}, \quad \mathbf{C}_2 = \begin{bmatrix} 3 \end{bmatrix}, \quad \mathbf{D}_2 = \begin{bmatrix} 0 \end{bmatrix}$$

P4.4 $G_1(z)$ and $G_2(z)$ given in TF form.

$$G_1(z) = \frac{0.75z^2 - 1.35z + 0.61}{z^2 - 1.62z + 0.73} \quad \text{and} \quad G_2(z) = \frac{0.38z + 0.33}{z^2 - 1.14z + 0.61}$$

Do this problem by first converting $G_1(z)$ and $G_2(z)$ to SS form and then performing the interconnections in SS form.

P4.5 Feedback connection of $G_1(z)$ and $G_2(z)$. Redo the feedback connection of the systems in Problem 4.4 by first performing the feedback connection in TF form and then converting the resulting system to SS form. You will obtain a state-space model with matrices **A**, **B**, and **C** different from those of the corresponding model obtained in Problem 4.4. In Problem 4.8, you will be asked to show that these two models are equivalent in the sense that they both have the same transfer function.

POLES, ZEROS, EIGENVALUES, AND STABILITY

The poles and zeros of a single-input/single-output (SISO) linear, time-invariant system in state-space form (4.1) can be computed directly from the SS form. This is done by using the Control System Toolbox command `[z,p,k] = zpkdata(G,'v')`, where the second input argument `'v'` forces the outputs `z` and `p` to appear as column vectors. The column vectors `z` and `p` contain the zeros and poles of the transfer function, respectively, and the scalar `k` contains the gain. For a multi-input system, the form of the command is `[zz,pp,kk] = zpkdata(G)`, where the outputs `zz` and `pp` are *cell arrays*, and `kk` is a matrix. As was explained in Example 2.12, the expressions `zz{:}` and `pp{:}` must be entered to display the numerical values of the zeros and poles.

Alternatively, the poles can be computed by using either the `pole(G)` or `eig(A)` command. For a discrete-time system, if all the eigenvalues of **A**, which are the system's poles, are inside the unit circle, the system is asymptotically stable.

For a SISO system, the zeros obtained from the ZPK form can also be computed by the transmission-zero command `tzero(G)` from the Control System Toolbox. A plot of the poles and zeros can be obtained directly from the `pzmap(G)` command. For high-order systems, the computation of the poles and zeros from the state-space matrices is numerically more reliable than solving for the roots of the numerator and denominator polynomials of a transfer function. The following example illustrates the use of the `ss`, `zpkdata`, `pole`, `tzero`, and `pzmap` functions.

☐ EXAMPLE 4.3 *Eigenvalues and Transmission Zeros*

Consider the system whose state-space representation is

$$\mathbf{A} = \begin{bmatrix} 0 & 1 & 0 \\ -0.65 & -0.7 & 1 \\ 0.25 & 0 & 0.5 \end{bmatrix}, \quad \mathbf{B} = \begin{bmatrix} 0 \\ 0 \\ 10 \end{bmatrix}$$

$$\mathbf{C} = \begin{bmatrix} 1 & 0 & 1 \end{bmatrix}, \quad \mathbf{D} = \begin{bmatrix} 0 \end{bmatrix}$$

and whose sampling period is 1 s. Build the system as a SS object and then obtain the zeros, poles, and gain of its transfer function. Verify that the resulting poles and zeros agree with those found by using the `pole` and `tzero` functions. Plot the zeros and poles on the z-plane and comment on the stability of the system.

Solution

Following the MATLAB commands in Script 4.3, we implement the state-space model as an SS object and extract its zeros, poles, and gain. Then the poles and zeros from the `zpkdata` command are expressed in polar form, yielding a real pole at $z = 0.6613$, a pair of complex poles at $z = 0.9325\epsilon^{\pm j2.0508}$, and a pair of complex zeros at $z = 1.2845\epsilon^{\pm j1.8468}$. The gain is 10. The use of the `pole` and `tzero` commands yields the same results.

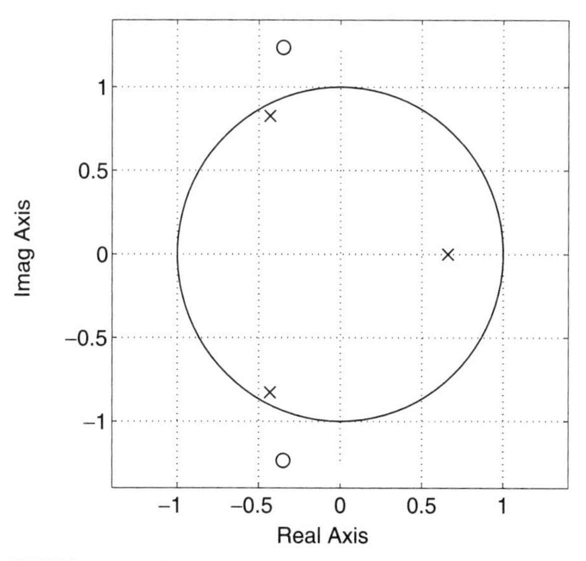

FIGURE 4.1 *Pole-zero plot for Example 4.3*

To plot the poles and zeros, we use the **pzmap** function with the SS object G as the single argument. Because all the poles of the system are inside the unit circle, the system is asymptotically stable. The fact that the two zeros are outside the unit circle has no implication regarding the stability of the system.

MATLAB Script

```
% Script 4.3:  Eigenvalues and transmission zeros
Ts = 1                                       % sampling period (s)
A = [0 1 0; -.65 -0.7 1; 0.25 0 0.5]         % state-space matrices
B = [0; 0; 10], C = [1 0 1], D = 0
G = ss(A,B,C,D,Ts)                           % build system as SS object
%--- extract zeros, poles, and gain from SS object
[zG,pG,kG] = zpkdata(G,'v')
%--- convert zeros to polar form and display
[mag_zG,theta_zG] = xy2p(zG)                 % zeros in polar form
%--- convert poles to polar form and display
[mag_pG,theta_pG] = xy2p(pG)                 % poles in polar form
%--- compute zeros using tzero command and convert to polar form
[mag_zeroG,theta_zeroG] = xy2p(tzero(G))     % zeros in polar form
%--- compute poles using pole command and convert to polar form
[mag_poleG,theta_poleG] = xy2p(pole(G))      % poles in polar form
dcgain(G)                                    % compute the DC gain
ucircle                                      % draw the unit circle
axis equal, hold on
pzmap(G), hold off                           % plot zeros & poles in z-plane
```

□

REINFORCEMENT PROBLEMS

In Problems 4.6 and 4.7, convert the state-space model to its transfer function in ZPK form; that is, find its zeros, poles, and gain. Then use the `pole` and `tzero` functions to verify the poles and zeros obtained by using the `zpkdata` command. Plot the poles and zeros on the z-plane.

P4.6 Third-order system. Use the state-space model from Problem 4.1.

P4.7 Fourth-order system with $\mathbf{D} \neq 0$. Use the state-space model from Problem 4.2. Note that because $\mathbf{A}$ is upper-triangular, the eigenvalues of the $\mathbf{A}$ matrix are the entries on its main diagonal.

P4.8 Equivalence of state-space models. Use `zpkdata` to find the zeros, poles, and gain of the state-space models obtained from the feedback connections in Problems 4.4 and 4.5. You should find that they are equal, showing that the systems are equivalent even though their state matrices differ.

TIME RESPONSE

One advantage of a state-space model over a transfer-function model is that the initial values of the state variables can be directly included in the model as the nonzero initial condition vector $\mathbf{x}(0)$. In contrast, we implicitly assume zero-state conditions whenever we use either the TF or the ZPK form. The complete response of the state vector $\mathbf{x}(k)$ of the system represented by (4.1) with an arbitrary input $\mathbf{u}(k)$ and the initial condition $\mathbf{x}(0)$ is

$$\mathbf{x}(k) = \mathbf{A}^k\mathbf{x}(0) + \sum_{i=0}^{k-1} \mathbf{A}^{k-i-1}\mathbf{B}\mathbf{u}(i) \tag{4.2}$$

where $\mathbf{A}^k$ is called the *state transition matrix* and i is a dummy index of summation over the discrete-time periods.

The response consists of two parts. In the first part, the matrix power function $\mathbf{A}^k$ maps the initial state $\mathbf{x}(0)$ to the current state $\mathbf{x}(k)$ if the input $\mathbf{u}(k)$ is identically zero. The second part of the response is the discrete convolution of the state transition matrix $\mathbf{A}^k$ with the input $\mathbf{u}(k)$, the solution for which requires a summation of matrix expressions.

In Chapter 2, the `impulse`, `step`, and `lsim` commands were used to simulate the time responses of discrete-time TF and ZPK objects. These commands can also be used for the simulation of discrete-time SS objects. In particular, `lsim` allows the inclusion of a nonzero initial condition as an argument and the calculation of the states. For example, the command `[y,t,x] = lsim(G,u,t,x0)` requires that `t` be a column vector of time values sampled at the sampling period of the system, the input `u` have as many columns as the number of inputs and as many rows as `t`, and `x0` be a column vector of the initial conditions of the n state variables. The output response is in `y`. The state response is returned in `x`, which has n columns and as many rows as `t`.

Because the system's sampling period T_s is known (or assumed to be unity if a value has not been specified), the time variable t is redundant an input argument, and can be omitted or set to the empty matrix `[ ]`. In this case, the number of rows in the input `u` establishes the number of time points, and the sampling period T_s establishes its value for any discrete time k according to the relationship $t = kT_s$.

In the following example, we illustrate the use of the time-simulation functions `impulse`, `step`, and `lsim` for a state-space object.

☐ EXAMPLE 4.4 *Impulse, Step, and General Response*

Use MATLAB to compute and plot the impulse and step responses of the state-space model

$$\mathbf{A} = \begin{bmatrix} 0.8 & -0.7 & 0.6 \\ 1 & 0 & 0 \\ 0 & 1 & 0 \end{bmatrix}, \quad \mathbf{B} = \begin{bmatrix} 1 \\ 0 \\ 0 \end{bmatrix}$$

$$\mathbf{C} = \begin{bmatrix} 0 & 1.5 & 1 \end{bmatrix}, \quad \mathbf{D} = \begin{bmatrix} 0 \end{bmatrix}$$

over the interval $0 \leq t \leq 50$ s, where the sampling period is $T_s = 1$ s. Then simulate and plot the zero-state response over the same time interval when the input is the piecewise constant (PWC) function

$$u(kT_s) = \begin{cases} 2, & 0 \leq kT_s < 20 \\ 0.5, & kT_s \geq 20 \end{cases}$$

Finally, simulate the system's response to this same PWC input, but with the nonzero initial state $\mathbf{x}(0) = [\, 1 \;\; 0 \;\; 2 \,]^T$.

Solution

The MATLAB commands in Script 4.4 create the system as an LTI object in state-space form, with a sampling period of unity. Then the impulse and step responses, which assume zero initial conditions, are produced. The resulting plots are shown in parts (a) and (b) of Figure 4.2, respectively. Note in Script 4.4 that these responses were obtained by using the `impulse` and `step` commands, respectively. These are the same commands used for continuous-time systems. However, these Control System Toolbox functions make use of the fact that the LTI object `G` represents a discrete-time system to evoke the appropriate simulation algorithm.

Next, the specified PWC input is defined over the interval $0 \leq t \leq 50$ by first defining it to have the value of 2 over the entire interval and then changing to 0.5 those points for which $t \geq 20$. Note that the third input argument in the `lsim` command is the empty matrix `[ ]`. Normally, this argument would specify the time vector. In this case, however, this information is already known from the sampling period T_s, which is a property of the LTI object, and the length of the input vector $u(kT_s)$, which specifies the number of sample time points.

The response of the output $y(kT_s)$ with zero initial state is obtained in part (c) of the script and plotted in Figure 4.2(c). Observe that for $0 \leq k \leq 21$, the response is exactly twice that of the step response, shown in Figure 4.2(b).

The response to the same PWC input, but with the nonzero initial state, is computed in part (d) of the script, again using the `lsim` command. The only change is that $\mathbf{x}(0)$ must be specified. The response is shown in Figure 4.2(d). Because the system is stable, the initial condition affects the response only up to about $k = 15$.

MATLAB Script

```
% Script 4.4:  Impulse, step, zero-state, and general responses
Ts = 1                                  % sampling period (s)
A = [0.8 -0.7 0.6; 1 0 0; 0 1 0]        % state-space matrices
B = [1; 0; 0]
C = [0 1.5 1]
D = 0
G = ss(A,B,C,D,Ts)                      % build system as discrete SS object
dt = [0:Ts:50];                         % discrete-time vector
%---------- (a) impulse response --------------
yi_G = impulse(G,dt);                   % use time vector dt
stem(dt,yi_G,':*'), grid                % plot with dotted stems
%---------- (b) step response --------------
ys_G = step(G,dt);                      % use time vector dt
stem(dt,ys_G,':*'), grid                % plot with dotted stems
%---------- (c) zero-state response to PWC input ----
u = 2*ones(size(dt));                   % start with all points = 2.0
i = find(dt>=20);                       % change values to 0.5
u(i) = 0.5;                             % ...for dt >= 20
x0 = [0; 0; 0]                          % zero IC
[y_G,t] = lsim(G,u,[ ],x0);             % response to input, with IC = 0
plot(t,y_G,'*'),grid,hold on            % plot output with *'s, then
plot(t,u,'o'),hold off                  % ...add input with o's
%---------- (d) response to PWC input and nonzero IC ----
x0 = [1; 0; 2]                          % nonzero initial state
[y_G,t] = lsim(G,u,[ ],x0);             % response to input and IC
plot(t,y_G,'*'),grid,hold on            % plot output, then
plot(t,u,'o'),hold off                  % ...add input
```

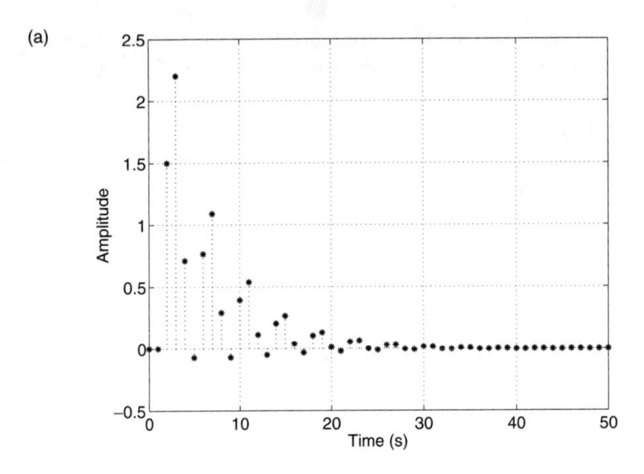
(a)
Amplitude
Time (s)

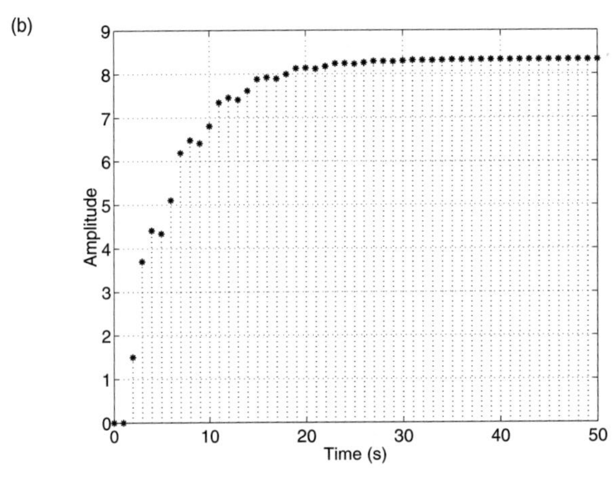
(b)
Amplitude
Time (s)

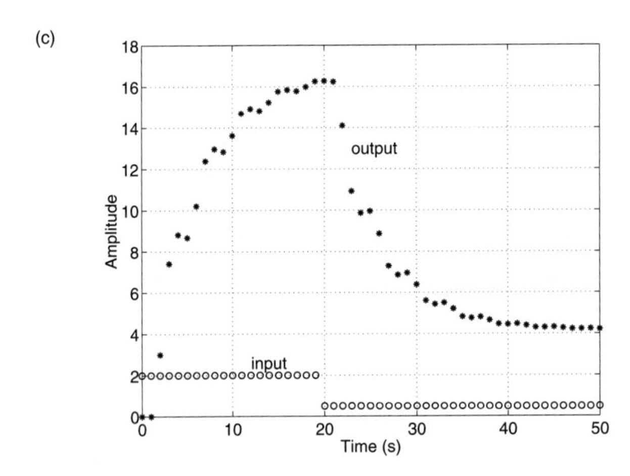
(c)
output
input
Amplitude
Time (s)

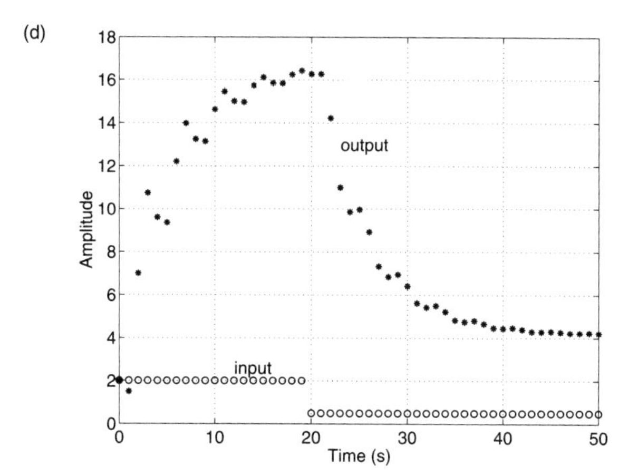

FIGURE 4.2 *Responses to various inputs for Example 4.4: (a) impulse; (b) step; (c) PWC input with zero initial state; (d) PWC input with nonzero initial state*

□

REINFORCEMENT PROBLEMS

P4.9 Third-order system. Simulate and plot the impulse and step responses of the state-space model

$$\mathbf{A} = \begin{bmatrix} -0.027 & -3.42 & -5.07 \\ 0.085 & 0.56 & -0.71 \\ 0.012 & 0.17 & 0.94 \end{bmatrix}, \quad \mathbf{B} = \begin{bmatrix} 0.85 \\ 0.12 \\ 0 \end{bmatrix}$$

$$\mathbf{C} = \begin{bmatrix} 2 & 7 & 5 \end{bmatrix}, \quad \mathbf{D} = \begin{bmatrix} 0 \end{bmatrix}$$

The sampling period is $T_s = 0.4$ s.

P4.10 General response. Simulate the responses of the three elements of the state vector $\mathbf{x}(k)$ and the scalar output $y(k)$ of the system in Problem 4.9 with a step input of amplitude 2 and the initial condition $\mathbf{x}(0) = [2\ 0\ 1]^T$. Plot the responses for $0 \le t \le 6$ s.

P4.11 Initial-condition responses. Solve for and plot the zero-input responses for the fourth-order system of Problem 4.2 with the initial condition of unity for one state variable at a time—namely, $\mathbf{x}(0) = [1\ 0\ 0\ 0]^T$, then $\mathbf{x}(0) = [0\ 1\ 0\ 0]^T$, etc.

To reveal additional system properties, the state-space model (4.1) can be transformed to a new set of state variables represented by the vector $\mathbf{z}$. Let $\mathbf{T}$ be the nonsingular transformation matrix relating $\mathbf{x}(k) = \mathbf{Tz}(k)$. In the new state vector $\mathbf{z}(k)$, model (4.1) becomes

$$\mathbf{z}(k+1) = \mathbf{T}^{-1}\mathbf{ATz}(k) + \mathbf{T}^{-1}\mathbf{Bu}(k), \quad \mathbf{y}(k) = \mathbf{CTz}(k) + \mathbf{Du}(k) \quad (4.3)$$

Models (4.1) and (4.3) are equivalent in the sense that they have the same input-output transfer function $\mathbf{G}(z)$ from $\mathbf{U}(z)$ to $\mathbf{Y}(z)$.

A particularly useful transformation happens when $\mathbf{T} = \mathbf{M}$, a matrix whose columns are the eigenvectors of $\mathbf{A}$. When $\mathbf{A}$ has distinct eigenvalues, $\mathbf{A}_d = \mathbf{M}^{-1}\mathbf{AM}$ is a diagonal matrix, and the resulting model (4.3) is said to be in *modal form*. The diagonal matrix $\mathbf{A}_d$ and the $\mathbf{M}$ matrix can be obtained using the MATLAB eigenvalue command `[M,Ad] = eig(A)` with two output variables. The ith column of $\mathbf{M}$ is the eigenvector corresponding to the ith eigenvalue of the diagonal matrix $\mathbf{A}_d$. The modal form is of interest because it has a parallel structure of first-order subsystems, and its response is in terms of the mode functions of the system. The Control System Toolbox `canon` function provides an option for finding the modal form directly, as shown in the following example.

☐ EXAMPLE 4.5 *Modal Transformation*

For the state-space model in Example 4.4, find the eigenvalues and eigenvectors of $\mathbf{A}$ and the transformed matrices $\mathbf{A}_d$, $\mathbf{B}_d$, and $\mathbf{C}_d$. Then use the `canon` command to find the modal form directly.

Solution

The MATLAB commands that perform the computations required in this example are given in Script 4.5. After entering the state matrices, the command `[M,Ad] = eig(A)` yields the eigenvector matrix

$$\mathbf{M} = \begin{bmatrix} 0.4675 & 0.4825 + j0.0164 & 0.4825 - j0.0164 \\ 0.5641 & 0.0096 - j0.5674 & 0.0096 + j0.5674 \\ 0.6806 & -0.6670 & -0.6670 \end{bmatrix}$$

and the diagonal matrix

$$\mathbf{A}_d = \begin{bmatrix} 0.8288 & 0 & 0 \\ 0 & -0.0144 + j0.8507 & 0 \\ 0 & 0 & -0.0144 - j0.8507 \end{bmatrix}$$

whose main-diagonal elements are the three eigenvalues of $\mathbf{A}$. Note that $\mathbf{A}_d$ has a conjugate pair of complex eigenvalues, and as a result the corresponding eigenvectors in $\mathbf{M}$ are also complex. The transformed input matrix is

$$\mathbf{B}_d = \mathbf{M}^{-1}\mathbf{B} = \begin{bmatrix} 1.0241 & 0.5225 - j0.5179 & 0.5225 + j0.5179 \end{bmatrix}^T$$

and the transformed output matrix is

$$\mathbf{C}_d = \mathbf{CM} = \begin{bmatrix} 1.5267 & -0.6525 - j0.8511 & -0.6525 + j0.8511 \end{bmatrix}$$

Both are complex. When we apply the `canon` function, in which the second argument specifies the modal transformation, we obtain the *real* modal form (no complex numbers) of the system

$$\mathbf{A}_m = \begin{bmatrix} 0.8288 & 0 & 0 \\ 0 & -0.0144 & 0.8507 \\ 0 & -0.8507 & -0.0144 \end{bmatrix}, \quad \mathbf{B}_m = \begin{bmatrix} 1.0241 \\ 1.0450 \\ 1.0358 \end{bmatrix}$$

$$\mathbf{C}_m = \begin{bmatrix} 1.5267 & -0.6525 & -0.8511 \end{bmatrix}, \quad \mathbf{D}_m = \begin{bmatrix} 0 \end{bmatrix}$$

When matrix $\mathbf{A}$ has complex eigenvalues, the `canon` function implements the modal form with real numbers only. The resulting $\mathbf{A}_m$ has 2×2 diagonal blocks instead of complex eigenvalues on the diagonals. The `eig` function can be applied to the lower 2×2 diagonal block of $\mathbf{A}_m$ to show that its eigenvalues are $-0.0144 \pm j0.8507 = 0.8508\epsilon^{\pm j1.5877}$.

In closing, we caution the reader that some of the matrices shown in Example 4.5 may not agree with the results obtained using a different version of the Control System Toolbox. The results shown here were obtained using version 6.0, which is part of MATLAB release 12. Different results were obtained for the second and third columns of $\mathbf{M}$, the second and third rows of $\mathbf{B}_d$ and $\mathbf{B}_m$, and the second and third columns of $\mathbf{C}_d$ and $\mathbf{C}_m$ when version 4 was used. The reason for these differences is that different forms were used to improve the numerical properties of the results. In any event, the behavior of the two systems from input to output will be the same. It is only the internal representation that may differ.

MATLAB Script

```
% Script 4.5:  Modal transformation
Ts = 1                               % sampling interval
A = [0.8 -0.7 0.6; 1 0 0; 0 1 0]   % state-space matrices
B = [1; 0; 0]
C = [0 1.5 1]
D = 0
G = ss(A,B,C,D,Ts)          % build system as SS object
[M,Ad] = eig(A)             % compute eigenvectors & diagonalized A matrix
Bd = inv(M)*B                       % transformed input matrix
Cd = C*M                            % transformed output matrix
Gm = canon(G,'modal')               % create real modal form as SS object
%------ compute eigenvalues of lower 2x2 block of Am ------
xy2p(eig(Am(2:3,2:3)))              % display in polar form
```

□

REINFORCEMENT PROBLEMS

For the state-space models given in the following problems, find the eigenvalues and eigenvectors of **A** and the modal form of the system. If the eigenvalues are complex, use the `canon` command to express the system in its real modal form.

P4.12 Third-order system. Use the state-space model from Problem 4.1.

P4.13 Fourth-order system. Use the state-space model from Problem 4.2.

EXPLORATORY PROBLEMS

EP4.1 Interactive analysis with MATLAB. The MATLAB file `done_blk.m` can be used to perform a variety of analytical tasks on a single-block, discrete-time system implemented as an SS object. You can create MATLAB models of the real-world systems described in Appendix A by running the appropriate M-files and selecting the state-space model on the menus. At the conclusion of the M-file, the state-space matrices will be in the workspace, with names like `Ad`, `Bd`, `Cd`, and `Dd`. You can find the eigenvalues and zeros of the model and examine time responses due to an impulse, a step, and a general input. You can create the transfer function of a proportional, lead, lag, or lead-lag controller and convert it to a state-space model. Then use the `*` operator and `feedback` command to build a closed-loop state-space model in the form of Figure 3.6. The `done_blk` program can then be used to analyze the resulting closed-loop system.

COMPREHENSIVE PROBLEMS

CP4.1 Ball and beam system. Run the file `dbbeam.m` to obtain the state-space model from the input voltage to the wheel angle θ and the ball position ξ. Examine the structure of the state-space matrices—that is, the locations of the zero and nonzero entries. Because the servo motor mechanism drives the ball dynamics, the **A** matrix has an upper-triangular structure with the chosen state ordering. Find the poles and zeros of the state-space model.

Implement the proportional controller $K_m(z)$ and the PD controller $K_b(z)$ designed in Comprehensive Problem CP3.1 in state-space form and close the loops sequentially, as in CP3.1. Examine the resulting state-space matrices, in particular, the zero entries that have become nonzero in the closed-loop **A** matrix. Note that the number of states has increased, because the dynamics of the PD controller are also included in the closed-loop system **A** matrix. Find the poles and zeros for each of the outputs of the closed-loop state-space model. Perform a closed-loop system response for a ball position command of 0.1 m. Next, setting the input to zero, apply only the initial condition $[\xi\ \dot{\xi}\ \theta\ \omega]^T = [0.1\ 0\ 0\ 0]^T$ to the system and simulate the time response using `lsim`. You will notice that the control system returns the ball to the origin from a position of 0.1 m away.

CP4.2 Inverted pendulum. Run the file `dstick.m` to generate the four matrices of the state-space model of the inverted pendulum in the form of (4.1) and verify that the dimensions of each matrix are consistent with the number of state variables (4), the number of outputs (2), and the number of inputs (1). Then use the `eig` command to find the eigenvalues of the open-loop system. (You should find one outside the unit circle, one at $z = 1$, and the other two on the real axis between $z = 0$ and $z = 1$.)

Use the `lsim` function with nonzero initial conditions to simulate the zero-input response to a *positive* initial pendulum angle $\theta(0) = 0.1$ rad and a *negative* initial angular velocity $\omega(0) = \omega_0$. Run the simulation a number of times with different values of ω_0 to find the initial angular velocity ω_0^* that causes the stick to approach the vertical position without going over the top or falling back. Because this is an unstable equilibrium condition, you will not be able to balance the stick for an indefinite period of time, but you should be able to satisfy the condition $|\theta| < 0.005$ rad at 1 s after a few attempts.

Use the `canon` function to transform the state-space model to modal form. Identify the eigenvalues in the modal **A** matrix and compare them with those found in the original **A** matrix with the `eig` function.

CP4.3 Electric power generation system. Run the file `dpower.m` to obtain the state-space model from the input (V_{field}) to each of the three outputs (V_{term}, ω, and P). Use the `pole` function to find the poles of the system and the `tzero` function to find the zeros for each output. Transform the system to the modal form and relate the resulting model to the poles already computed.

Consider the voltage control of the electric power system shown in Figure A.8 in Appendix A using a PI controller whose transfer function is $K_V(z) = K_P[(1+K_I T_s)z-1]/(z-1)$. Select the integral gain K_I such that the zero of the controller is at $z = 0.9$, and vary the proportional gain K_P from 10 to 50 in increments of 10 to obtain $K_V(z)$ in SS form. For each value of K_P, determine the poles, zeros, and DC gain of the closed-loop system, and plot its response to a step of amplitude 0.05 in the reference voltage input V_{ref}.

CP4.4 Hydroturbine system. Run the file `dhydro.m` to obtain the state-space model from the input u to the output power P of the hydroturbine system. Use the `pole` function to find the poles of the system and the `tzero` function to find the zeros. Obtain the model in the modal form and examine the resulting matrices. Design a PI controller for the hydroturbine system using the state-space form, as outlined in Comprehensive Problem CP3.4c. Perform a step response and observe the output response.

SUMMARY

In this chapter we have shown how to create and work with state-space models in MATLAB, and we have illustrated the computation of the poles and zeros of state-space models. The simulation and interconnection functions introduced in Chapters 2 and 3 for transfer-function models were extended to state-space models. We also examined the transformation of a state-space model to its modal form. In Chapter 10, we show how MATLAB can be used to design controllers for state-space models.

ANSWERS

P4.1 Transfer function is
$G(z) = (0.58z^2 - 0.1638z - 0.1946)/(z^3 - 0.96z^2 - 0.2395z + 0.257)$.
State-space matrices from TF form are

$$\mathbf{A}_1 = \begin{bmatrix} 0.96 & 0.2395 & -0.2570 \\ 1 & 0 & 0 \\ 0 & 1 & 0 \end{bmatrix}, \quad \mathbf{B}_1 = \begin{bmatrix} 1 \\ 0 \\ 0 \end{bmatrix}$$

$$\mathbf{C}_1 = \begin{bmatrix} 0.58 & -0.1638 & -0.1946 \end{bmatrix}, \quad \mathbf{D}_1 = \begin{bmatrix} 0 \end{bmatrix}$$

P4.2 Transfer function is $G(z) = (0.5z^4 - 0.2564z^3 - 0.1109z^2 + 0.218z - 0.1121)/(z^4 - 0.825z^3 - 0.4708z^2 + 0.5677z - 0.1224)$. State-space form is

$$\mathbf{A}_1 = \begin{bmatrix} 0.8250 & 0.4708 & -0.5677 & 0.2447 \\ 1 & 0 & 0 & 0 \\ 0 & 1 & 0 & 0 \\ 0 & 0 & 0.5 & 0 \end{bmatrix}, \quad \mathbf{B}_1 = \begin{bmatrix} 1 \\ 0 \\ 0 \\ 0 \end{bmatrix}$$

$$\mathbf{C}_1 = \begin{bmatrix} 0.1561 & 0.1246 & -0.0658 & -0.1019 \end{bmatrix}, \quad \mathbf{D}_1 = \begin{bmatrix} 0.5 \end{bmatrix}$$

P4.3 Feedback connection:

$$\mathbf{A}_f = \begin{bmatrix} 0.98 & 0.18 & 0.014 & -0.003 \\ -0.22 & 0.71 & 0.11 & -0.042 \\ -1.76 & -2.42 & 0.16 & -0.33 \\ 0.4 & 0.2 & 0 & 0.2 \end{bmatrix}, \quad \mathbf{B}_f = \begin{bmatrix} 0.001 \\ 0.014 \\ 0.11 \\ 0 \end{bmatrix}$$

$$\mathbf{C}_f = \begin{bmatrix} 4 & 2 & 0 & 0 \end{bmatrix}, \quad \mathbf{D}_f = \begin{bmatrix} 0 \end{bmatrix}$$

P4.4 Feedback connection:

$$\mathbf{A}_f = \begin{bmatrix} 1.62 & -0.73 & -0.19 & -0.165 \\ 1 & 0 & 0 & 0 \\ -0.27 & 0.125 & 0.855 & -0.8575 \\ 0 & 0 & 1 & 0 \end{bmatrix}, \quad \mathbf{B}_f = \begin{bmatrix} 0.5 \\ 0 \\ 0.75 \\ 0 \end{bmatrix}$$

$$\mathbf{C}_f = \begin{bmatrix} -0.27 & 0.125 & -0.285 & -0.2475 \end{bmatrix}, \quad \mathbf{D}_f = \begin{bmatrix} 0.75 \end{bmatrix}$$

P4.5 Feedback connection:

$$\mathbf{A}_f = \begin{bmatrix} 2.475 & -1.4607 & 1.0171 & -0.6466 \\ 2 & 0 & 0 & 0 \\ 0 & 1 & 0 & 0 \\ 0 & 0 & 0.5 & 0 \end{bmatrix}, \quad \mathbf{B}_f = \begin{bmatrix} 1 \\ 0 \\ 0 \\ 0 \end{bmatrix}$$

$$\mathbf{C}_f = \begin{bmatrix} -0.3488 & 0.2078 & 0.003338 & -0.1129 \end{bmatrix}, \quad \mathbf{D}_f = \begin{bmatrix} 0.75 \end{bmatrix}$$

P4.6 Zeros are at $z = -0.4550$ and 0.7374; poles are at $z = -0.5079, 0.5531,$ and 0.9147; gain is 0.5800.

P4.7 Zeros are at $z = -0.8184, 0.6446,$ and $0.6520\epsilon^{\pm j1.0163}$; poles are at $z = -0.8190, 0.3680, 0.6060,$ and 0.6700; gain is 0.5.

P4.8 Zeros are at $z = 0.7810\epsilon^{\pm j0.7528}$ and $0.9018\epsilon^{\pm j0.0641}$; poles are at $z = 0.8727\epsilon^{\pm j0.2096}$ and $0.9214\epsilon^{\pm j1.1411}$; gain is 0.75.

P4.9 Impulse response has a maximum of 2.540 at $t = 0.4$ s and settles to 0. Step response has a maximum of 2.802 at $t = 0.8$ s and settles to 1.0.

P4.10 Output $y(k)$ has a maximum of 9.0 at $t = 0$, a minimum of -6.392 at $t = 0.8$ s, and a value of 2.044 at $t = 6$ s.

P4.11 For $\mathbf{x}(0) = [1\ 0\ 0\ 0]^{\mathrm{T}}, y(k)$ has a maximum of 1.0 at $t = 0$ and a minimum of -0.819 at $t = 0.5$, and it settles to 0 as k increases.

P4.12 Diagonal form of the system matrix is

$$\mathbf{A}_d = \mathrm{diag}(-0.5079, 0.5531, 0.9147)$$

Real modal form is

$$\mathbf{A}_m = \begin{bmatrix} -0.5079 & 0 & 0 \\ 0 & 0.5531 & 0 \\ 0 & 0 & 0.9147 \end{bmatrix}, \quad \mathbf{B}_m = \begin{bmatrix} 0.3129 \\ -0.7577 \\ -0.5161 \end{bmatrix}$$

$$\mathbf{C}_m = \begin{bmatrix} 0.0808 & -0.3707 & -0.5306 \end{bmatrix}, \quad \mathbf{D}_m = \begin{bmatrix} 0 \end{bmatrix}$$

P4.13 Diagonal form of the system matrix is

$$\mathbf{A}_d = \text{diag}(-0.819, 0.670, 0.6060, 0.3680)$$

Modal form is

$$\mathbf{A}_m = \mathbf{A}_d, \quad \mathbf{B}_m = \begin{bmatrix} -0.0003 \\ 2.7485 \\ 3.2478 \\ 0.9891 \end{bmatrix}$$

$$\mathbf{C}_m = \begin{bmatrix} 1 & 0.0989 & 0.1468 & -0.5986 \end{bmatrix}, \quad \mathbf{D}_m = \begin{bmatrix} 0.5 \end{bmatrix}$$

5 *Sampled-Data Control Systems*

PREVIEW

With the improved computing capability and lower costs of microprocessors, it has become practical to use them to implement control designs. In addition, complex control logic is easier to implement on digital controllers than analog controllers. A system in which a digital controller drives a continuous plant is called a *sampled-data control system.* A sampler and a hold circuit serve as the interface between the continuous-time system and the digital system. In this chapter, we provide the links between continuous-time and discrete-time signals—namely, the processes of sampling and reconstruction. The concept of impulse sampling is illustrated using MATLAB. A key concept in sampling, *aliasing,* is shown in MATLAB plots in the time domain as well as in the frequency domain. In reconstruction, we study the operation of a zero-order hold (ZOH). Then we discuss the discretization of continuous-time systems driven by zero-order holds. Examples using MATLAB, Control System Toolbox functions, and RPI functions are provided.

IMPULSE SAMPLING

Impulse sampling of a continuous-time signal $e(t)$ is modeled by multiplying an impulse modulator at a uniform sampling period of T_s, as shown in Figure 5.1(a). Symbolically, the multiplication is represented as a switch, as shown in Figure 5.1(b). The mathematical expression for the impulse modulator is

$$\delta_{T_s}(t) = \delta(t) + \delta(t - T_s) + \delta(t - 2T_s) + \cdots = \sum_{k=0}^{\infty} \delta(t - kT_s) \tag{5.1}$$

where $\delta(t)$ is the Dirac function.

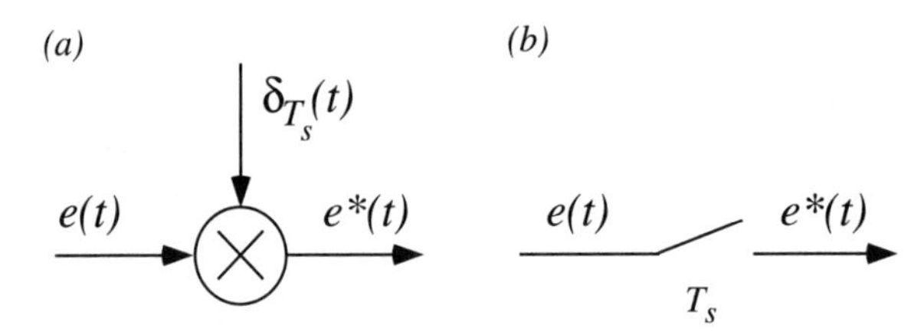

FIGURE 5.1 *Impulse sampling*

The resulting sampled signal, denoted by the superscript $*$, is given by

$$\begin{aligned} e^*(t) &= e(t)\delta_{T_s}(t) \\ &= e(t)\delta(t) + e(t)\delta(t - T_s) + e(t)\delta(t - 2T_s) + \cdots \\ &= e(0)\delta(t) + e(T_s)\delta(t - T_s) + e(2T_s)\delta(t - 2T_s) + \cdots \\ &= \sum_{k=0}^{\infty} e(kT_s)\delta(t - kT_s) \end{aligned}$$

Note that $e^*(t)$ is a continuous-time signal, although it is nonzero only at the time instants kT_s.

In practice, sampling is usually performed by an analog-to-digital (A/D) converter, and the process is more accurately modeled as pulse sampling. However, it is beyond the scope of this text to deal with such details.

In the following example, we illustrate the use of the RPI function `dplot` to plot sampled signals.

☐ EXAMPLE 5.1 *Impulse Sampling*

The continuous signals $e_1(t)$ and $e_2(t)$, defined by the expressions

$$\begin{aligned} e_1(t) &= 2\epsilon^{-t} + t\epsilon^{-t}, \quad t \geq 0 \\ e_2(t) &= \sin(2\pi t) + 0.5\cos(4\pi t + 45^\circ), \quad t \geq 0 \end{aligned}$$

are sampled with an ideal sampler at a frequency of 10 Hz. Plot $e_1^*(t)$ and $e_2^*(t)$ over the interval $0 \le t \le 2$ s.

Solution

To find the sampled signals, we need to compute $e_i(kT_s)$, where k is an integer starting from 0 and $T_s = 1/10 = 0.1$ s. For convenience, we set up the column vector **kT** for the sampling times kT_s. Note that the symbols "`.*`" together mean *element-by-element* multiplication of two vectors, rather than the more conventional vector products. The RPI function **dplot** represents the sampled signals $e_i^*(t)$ with triangles to signify that the signals consist of impulses. The plots of $e_1^*(t)$ and $e_2^*(t)$ are shown in parts (a) and (b) of Figure 5.2.

MATLAB Script

```
% Script 5.1:  Impulse sampling
Ts = 0.1
t = [0:0.02:2]';                        % continuous time
kT = [0:Ts:2]'                          % sampling at 10 Hz
%------ signals e_1(t) and e_1(k*Ts) ---------------
e_1t = 2*exp(-t) + t.*exp(-t);          % continuous signal e_1(t)
e_1k = 2*exp(-kT) + kT.*exp(-kT)        % sampled signal e_1(k*Ts)
dplot(kT,e_1k),grid,hold on             % plot sampled signal
plot(t,e_1t,'--'),hold off              % plot continuous signal with dashes
%------ signals e_2(t) and e_2(k*Ts) ---------------
e_2t = sin(2*pi*t) + 0.5*cos(4*pi*t+45/180*pi);   % continuous signal e_2(t)
e_2k = sin(2*pi*kT) + 0.5*cos(4*pi*kT+45/180*pi)  % sampled signal e_2(k*Ts)
dplot(kT,e_2k),grid,hold on             % plot sampled signal
plot(t,e_2t,'--'),hold off              % plot continuous signal with dashes
```

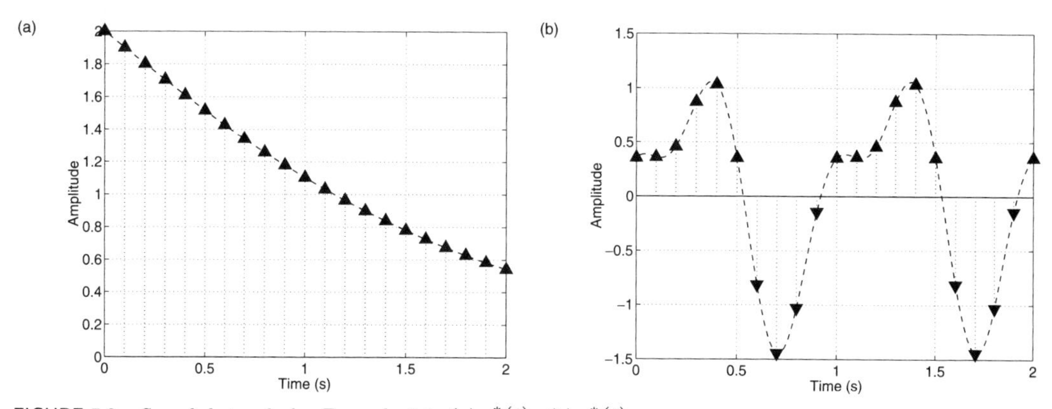

FIGURE 5.2 *Sampled signals for Example 5.1: (a) $e_1^*(t)$; (b) $e_2^*(t)$*

□

REINFORCEMENT PROBLEMS

For the continuous-time signals below, plot the signals obtained from impulse sampling for an appropriate interval of $t \geq 0$.

P5.1 Exponential signal.

$$e(t) = \epsilon^{-3t} + t\epsilon^{-2t} + 3\epsilon^{-10t} \quad \text{and} \quad T_s = 0.05 \text{ s}$$

P5.2 Sinusoidal signal.

$$e(t) = \epsilon^{-t}\cos(4\pi t) + 0.5\sin(10\pi t - 60^\circ) \quad \text{and} \quad T_s = 0.04 \text{ s}$$

ALIASING

When a continuous-time signal is sampled, the signal information between the sample points is not represented and actually may be lost. In such a situation, it is possible for different continuous signals to yield the *same* discrete-time signal. For example, a high-frequency continuous-time signal and a low-frequency continuous-time signal may yield the same sampled signal. Such a phenomenon is called *aliasing.* To avoid aliasing, the sampling frequency must be at least twice the highest frequency component in the continuous-time signal. This sampling frequency is known as the *Nyquist frequency.*

TIME-DOMAIN VIEW

The following example illustrates aliasing in the time domain.

□ EXAMPLE 5.2 *Aliasing in the Time Domain*

The following periodic continuous-time signals contain frequencies of 1, 9, and 11 Hz:

$$\begin{aligned} e_1(t) &= \sin(2\pi t) \\ e_2(t) &= \sin(18\pi t + 180^\circ) \\ e_3(t) &= \sin(22\pi t) \end{aligned}$$

Each signal is sampled at a frequency of 10 Hz to produce the signals $e_i^*(t)$, $i = 1, 2, 3$. Plot the sampled signals from $t = 0$ to 1 s on a single plot. Then plot all the continuous-time signals $e_i(t)$ on a separate plot.

Solution

The sampling of the three signals $e_1(t)$, $e_2(t)$ and $e_3(t)$ at 10 Hz is performed in Script 5.2, yielding the sampled signal shown in Figure 5.3(a). Hence, the 9-Hz and 11-Hz signals have been disguised as a 1-Hz signal. This is due to the fact that all three of the continuous-time signals have *identical* values at the sampling instants, marked by the $*$ symbols in Figure 5.3(b). Looking at the sampled signal in Figure 5.3(a), one is tempted to say that the frequency of the continuous-time signal must be 1 Hz. Because the human eye is conditioned to be a lowpass filter, the typical human reaction is to perform a smooth visual interpolation of the sampled signal.

MATLAB Script

```
% Script 5.2:  Aliasing in the time domain
fs = 10, Ts = 1/fs                   % sampling at 10 Hz
t = [0:Ts:1]'                        % time points from 0 to 1 sec
e_1 = sin(2*pi*t)                    % 1 Hz signal
e_2 = sin(18*pi*t+pi)                % 9 Hz signal
e_3 = sin(22*pi*t)                   % 11 Hz signal
%--------------- plot three sampled signals ---------
dplot(t,e_1)                         % plot the sampled 1 Hz signal
hold on
dplot(t,e_2)                         % plot the sampled 9 Hz signal
dplot(t,e_3)                         % plot the sampled 11 Hz signal
hold off
%-------------- plot three continuous-time signals ---------
tt = [0:0.01:1]';                    % consider tt as continuous time
plot(tt,sin(2*pi*tt),'-')            % solid curve
hold on
plot(tt,sin(18*pi*tt+pi),'--')       % dashed curve
plot(tt,sin(22*pi*tt),'-.')          % dash-dot curve
plot(t,e_1,'*')                      % mark intersections with *
hold off
```

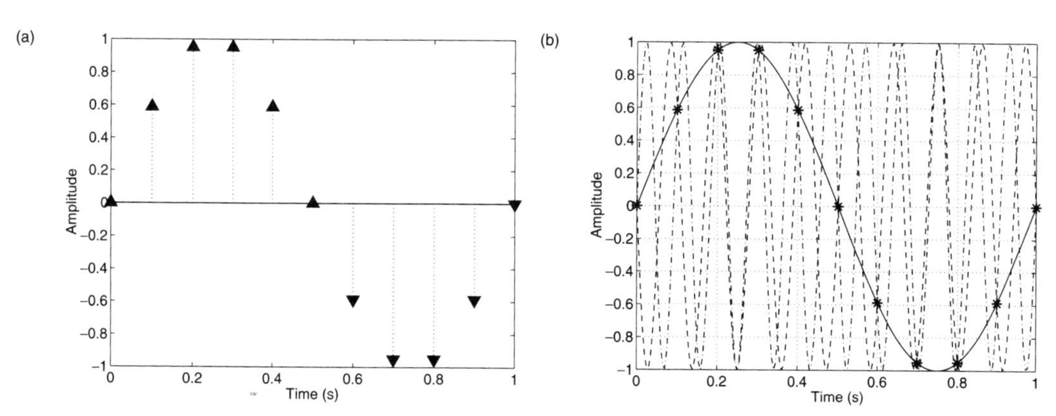

FIGURE 5.3 *Signals for Example 5.2: (a) sampled signals $e_1^*(t)$, $e_2^*(t)$ and $e_3^*(t)$; (b) continuous signals $e_1(t)$, $e_2(t)$ and $e_3(t)$*

WHAT IF? Redo Example 5.2 using a sampling frequency of 20 Hz. Do you still observe any aliasing effects? Repeat the analysis for sampling at 30 Hz. ■

□

FREQUENCY-DOMAIN VIEW

Aliasing can also be explained by examining the frequency content in a sampled signal. As shown by the signals in Example 5.2, the sampled signals have frequencies of 1, 9, and 11 Hz. More precisely, a continuous-time periodic signal with a frequency of f_o Hz sampled at f_s Hz will have frequency components at $\pm f_o \pm n f_s$ Hz, for $n = 0, 1, ..., \infty$. To compute the frequency spectrum of the signals, we can apply the fast Fourier transform command `fft` and use a paste operation to extend the frequencies to as large a value of n as desired. A brief discussion of fast Fourier transform is given in Appendix B. Example 5.3 illustrates this idea.

□ EXAMPLE 5.3 *Frequency Components in a Sampled Signal*

Plot the frequency spectrum of the signal obtained from sampling the continuous-time signal $e_1(t) = \sin(2\pi t)$ at a frequency of 10 Hz.

Solution

In Script 5.3, we use the same code as in Script 5.2 to generate the signal $e_1^*(t)$. Then the fast Fourier transform function `fft` is used to compute the frequency coefficients `fr`. Because $e_1(t)$ is an odd function, its Fourier transform is purely imaginary. Thus we extract the imaginary part of the spectrum by using the `imag` function. The `fr` coefficients are for frequencies from 0 to 9 Hz.

To continue beyond 9 Hz, we simply duplicate the spectrum by pasting as many `fr` arrays together as are desired. Here we paste three `fr` arrays together, expanding the frequency range from 0 to 29 Hz. The spectrum is shown in Figure 5.4. This plot clearly shows the additional frequency components when the signal is sampled. If the spectra of the sampled signals $e_2^*(t)$ and $e_3^*(t)$ were computed and plotted in a similar manner, they would be identical to the one in Figure 5.4.

MATLAB Script

```
% Script 5.3:  Spectrum of sampled signal
fs = 10, Ts = 1/fs              % sampling at 10 Hz
t = [0:Ts:0.9]'                 % time points from 0 to 0.9 sec
e_1 = sin(2*pi*t)               % 1-Hz signal
N = 10                          % length of data
fr = fft(e_1)/N                 % sampled signal spectrum
fr = imag(fr)                   % extract imaginary part
fr_paste = [fr; fr; fr]         % continue the spectrum
freq = [0:1:29]'                % extended frequency scale
dplot(freq,fr_paste)
```

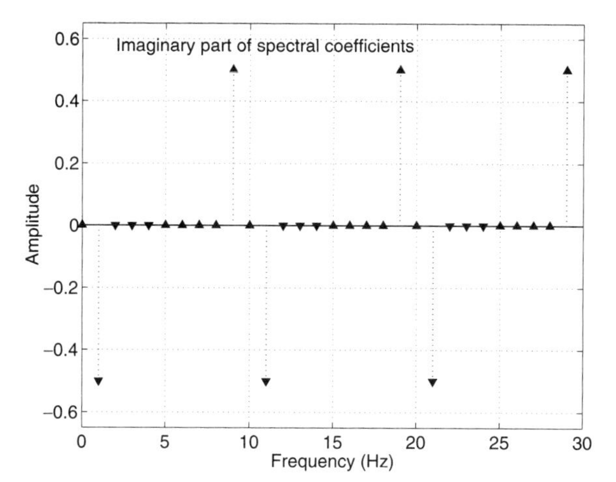

FIGURE 5.4 *Frequency spectrum of the sampled signal $e_1^*(t)$ for Example 5.3*

WHAT IF? Example 5.3 showed that when the signal $\sin(2\pi t)$ is sampled, the resulting spectrum is a set of imaginary values. The cosine signal $\cos(2\pi t)$ is the sine signal $\sin(2\pi t)$ shifted in phase by 90 degrees. Show that the spectrum of the cosine signal $\cos(2\pi t)$, when sampled at 10 Hz, has the same frequency content, but with real spectral coefficients. ■

□

REINFORCEMENT PROBLEMS

P5.3 Several periodic signals. Sample the following signals at 20 Hz and show that they result in the same sampled signal:

$$\begin{aligned} e_1(t) &= \sin(4\pi t + 20^\circ) + 2\cos(26\pi t + 45^\circ) \\ e_2(t) &= 2\cos(14\pi t - 45^\circ) + \cos(36\pi t + 70^\circ) \\ e_3(t) &= \sin(4\pi t + 20^\circ) - 2\cos(14\pi t + 135^\circ) \end{aligned}$$

In addition, plot the spectra of the sampled signals.

P5.4 Sampling at higher frequencies. Redo Problem 5.3, sampling the signals at 30 Hz and 40 Hz. Does aliasing occur in either case?

P5.5 Constant signals. The signals $e_1(t) = 1$ and $e_2(t) = \cos(100 \times 2\pi t)$ are sampled at 100 Hz. Show that the sampled signals are identical impulse trains with constant values. Then make up another signal that, when sampled, results in the same impulse train.

P5.6 Sampling at twice the highest frequency. Sample the signal $e_2(t)$ in Problem 5.5 at 200 Hz. Plot the sampled signal and its spectrum. Find a power function representation of $e_2^*(t)$.

ZERO-ORDER HOLD

Before a digital or discrete-time signal becomes the input to a continuous-time plant, it must be converted to a continuous-time signal that approximates it in some sense. Devices that perform this function are called *hold circuits.* According to Nyquist's ideal-sampling theorem, if a band-limited continuous-time signal $e(t)$ with frequencies no higher than ω_B is sampled at a frequency higher than $2\omega_B$, $e(t)$ can be recovered exactly by passing the sampled signal $e^*(t)$ through an ideal lowpass filter with a bandwidth of ω_B. Ideal lowpass filters are noncausal and thus cannot be realized, however.

In this section, we consider a commonly used hold function—namely, the *zero-order hold (ZOH)*—which is represented in Figure 5.5. In a digital control system, the digital-to-analog (D/A) converter performs the function of a ZOH. Exploratory Problem EP5.1 and Comprehensive Problem CP5.2 provide opportunities to work with the first-order hold (FOH) and fractional-order holds.

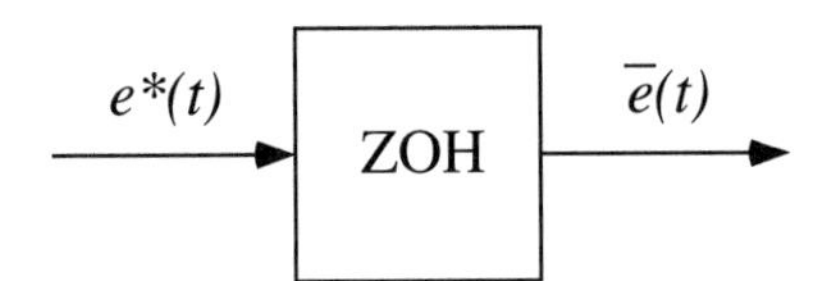

FIGURE 5.5 *Signal reconstruction using a zero-order hold*

When the sampled signal $e^*(t)$ is the input to a ZOH, the output is given by

$$\bar{e}(t) = e(kT_s), \quad kT_s \leq t < (k+1)T_s$$

Thus the impulse response of a ZOH is a pulse of unit height and width T_s that can be expressed analytically as

$$h_0(t) = U(t) - U(t - T_s) \tag{5.2}$$

where $U(t)$ is the unit-step function.

☐ EXAMPLE 5.4 *Output of a ZOH*

The 1-Hz continuous-time signal $e(t) = \sin(2\pi t)$ is sampled at a frequency of 10 Hz. Then the sampled signal $e^*(t)$ is passed through a ZOH. Find and plot the output $\bar{e}(t)$ of the ZOH from $t = 0$ to 1 s.

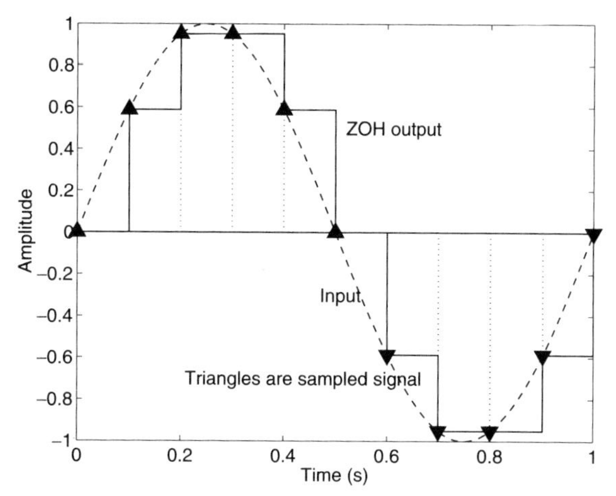

FIGURE 5.6 *Input signal (dashed), sampled signal (triangles), and output of ZOH (solid) in Example 5.4*

Solution

To determine the output of a ZOH, we use the MATLAB **stairs** function to plot the sampled input data points. Script 5.4 also includes the plotting of the original sine wave and the sampled sine wave. Figure 5.6 shows the output of the ZOH as well as $e(t)$ and $e^*(t)$.

MATLAB Script

```
% Script 5.4:  Output of a ZOH
fs = 10, Ts = 1/fs              % sampling frequency and period
dt = [0:Ts:1]'                  % time points from 0 to 1 s
e = sin(2*pi*dt)                % 1-Hz signal
dplot(dt,e), hold on            % sampled signal
stairs(dt,e)                    % ZOH output
tt = [0:0.01:1]';               % take tt as continuous time
ec = sin(2*pi*tt);              % 1-Hz signal
plot(tt,ec,'--'), hold off      % input sine wave
```

□

Figure 5.6 shows the discrepancy between the original continuous-time signal and the output of the ZOH. To better understand this difference, we need to investigate the frequency response characteristics of a ZOH. The transfer function of a ZOH is obtained by applying Laplace transform to (5.2), resulting in

$$G_{h0}(s) = \frac{1 - \epsilon^{-sT_s}}{s}$$

where ϵ^{-sT_s} is a delay of T_s in seconds. Substituting $j\omega$ for s, we see that the frequency response is

$$G_{h0}(j\omega) = \frac{1 - \epsilon^{-j\omega T_s}}{j\omega} \tag{5.3}$$

Because of the presence of $\epsilon^{-j\omega T_s}$, this function cannot be directly computed by the MATLAB frequency response `bode` function. In Example 5.5, we plot the frequency response of the ZOH. Then in Example 5.6, we use the ZOH frequency response to determine the spectrum of the ZOH output when its input is a sampled sine wave, and to illustrate the extent to which the time-domain signal consisting of the five lowest frequency components approximates the ZOH output.

☐ EXAMPLE 5.5 *Frequency Response of a ZOH*

Plot the magnitude and phase of the ZOH frequency response in the frequency range of 0 to 30 Hz. Use a sampling frequency of 10 Hz.

Solution

In Script 5.5, we first generate a frequency range of 0 to 30 Hz, and we then use (5.3) to generate the ZOH frequency response `Gh0`. Note that at DC, the denominator of (5.3) is zero. As a result, direct computation of (5.3) using MATLAB results in a divide-by-zero warning message. Using the L'Hôpital's rule, it can be shown that the DC gain is T_s, the sampling period. The ZOH frequency response is shown in Figure 5.7. The magnitude of the frequency response is a sinc function.[1]

MATLAB Script

```
% Script 5.5:  Frequency response of Gh0(s)
fs = 10, Ts = 1/fs                       % sampling frequency and period
f = [0:0.1:30]'                          % frequency points from 0 to 30 Hz
Gh0 = (1-exp(-j*Ts*2*pi*f))./(j*2*pi*f)  % './' is point-by-point divide
Gh0(1) = Ts                              % define DC gain separately
subplot(2,1,1), plot(f,abs(Gh0))         % magnitude plot
subplot(2,1,2), plot(f,angle(Gh0)*180/pi) % phase plot
```

☐

☐ EXAMPLE 5.6 *Frequency Response of the ZOH output*

Use the frequency response of the ZOH determined in Example 5.5 to find the frequency spectrum of the ZOH output $\bar{e}(t)$ that was considered in Example 5.4. Then plot the signal made up of the five lowest frequency components of $\bar{e}(t)$.

[1]The sinc function is defined as $\text{sinc}(x) = \sin(\pi x)/\pi x$. It can be shown that $G_{h0}(j\omega)$ in (5.3) can be simplified to $T_s \text{sinc}(\omega/\omega_s)\epsilon^{-j(\pi\omega/\omega_s)}$.

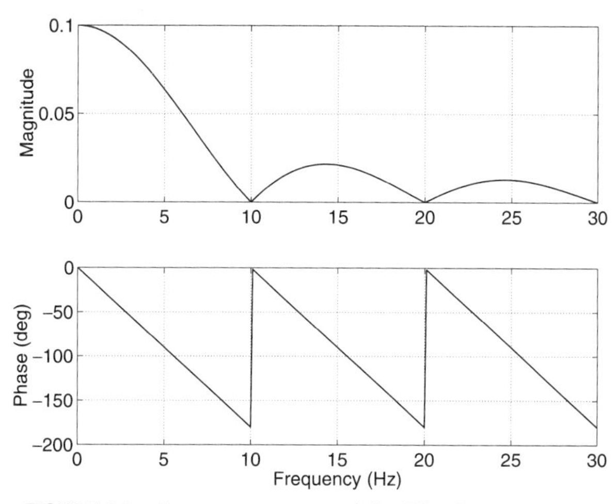

FIGURE 5.7 *Frequency response of the ZOH for Example 5.5*

Solution

To obtain the spectrum of the ZOH output, we first compute the spectrum of the sampled signal using the `fft` function and paste the results, as in Example 5.3. Here we set the zero coefficients in the spectrum to an "exact" zero using the RPI function `small20`. Then the spectrum of the ZOH output signal is obtained by

$$\bar{E}(j\omega) = G_{h0}(j\omega)E^*(j\omega)$$

where $E^*(j\omega)$ is the spectrum of $e^*(t)$. This is plotted in Figure 5.8(a). Finally, the nonzero frequency components α_i and α_{-i} corresponding to the frequencies $\pm f_i$ Hz generate the sinusoidal signal

$$\bar{e}_i(t) = 2|\alpha_i|\cos(2\pi f_i t + \arg(\alpha_i))$$

The five lowest frequency components in $\bar{e}(t)$ are 1, 9, 11, 19, and 21 Hz. For convenience, we program the summation of these five components using the `for` loop, which sums all the sinusoidal components up to and including 21 Hz, in the MATLAB script. In Figure 5.8(b), this approximated $\bar{e}(t)$ is plotted against $\bar{e}(t)$. Note that the approximation is adequate except for the ripples. This is known as the *Gibbs phenomenon.*

MATLAB Script

```
% Script 5.6:  Frequency response of the ZOH output
fs = 10, Ts = 1/fs                    % sampling frequency and period
f = [0:1:29]'                         % 30 frequency points from 0 to 29 Hz
Gh0 = (1-exp(-j*Ts*2*pi*f))./(j*2*pi*f) % './' is entry-by-entry divide
Gh0(1) = Ts                             % define DC gain separately
% spectrum of ZOH output
dt = [0:Ts:0.9]'                   % discrete time points from 0 to 0.9 s
e = sin(2*pi*dt)                   % 1 Hz signal
N = 10                             % number of data points
fr = fft(e)/N/Ts                   % frequency spectrum
```

cont.

```
fr = small20(fr,1e-10)          % set small coeff to exact zeros
fr_paste = [fr; fr; fr]         % continue the spectrum to form 30 freq pts
fr_e = fr_paste.*Gh0            % '.*' is entry-by-entry multiply
subplot(2,1,1), dplot(f,abs(fr_e))               % magnitude plot
subplot(2,1,2), stem(f,angle(fr_e)*180/pi,':')  % phase plot
% sinusoidal series approximation of ZOH output
tt = [0:0.001:1]';
ff = [1 9 11 19 21]             % first 5 nonzero frequency components
e5 = 0*tt;
for i = 1:5                     % sum over all 5 frequency components
  % 1 is added to ff(i) because the 1st entry of fr_e is DC
  e5=e5+2*abs(fr_e(ff(i)+1))*cos(2*pi*ff(i)*tt+angle(fr_e(ff(i)+1)));
end
plot(tt,e5), hold on, stairs(dt,e,'--'), hold off
```

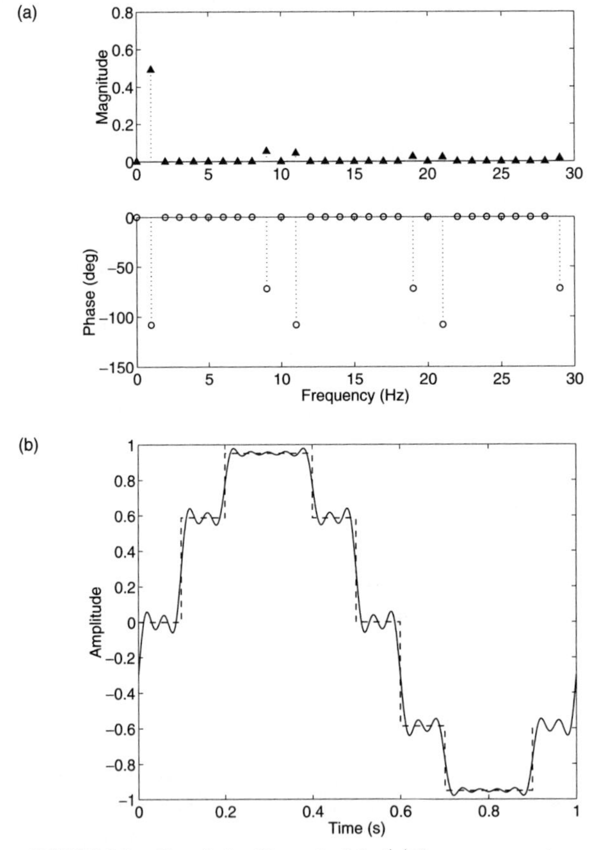

FIGURE 5.8 *Signals for Example 5.6: (a) frequency spectrum of $\bar{e}(t)$; (b) ZOH output (dashed) and signal made up of the five lowest frequency components of $\bar{e}(t)$ (solid)*

W H A T I F ? Expand Example 5.6 to obtain a signal consisting of the ten lowest frequency components of $\bar{e}(t)$. Do you still observe the Gibbs phenomenon? ■

□

REINFORCEMENT PROBLEMS

P5.7 Several periodic signals. Sample the signal $e(t) = \sin(4\pi t + 20°) + 2\cos(26\pi t + 45°)$ at 20 Hz and reconstruct a continuous-time signal using a ZOH. In addition, plot the frequency spectrum of the ZOH output.

P5.8 Triangular wave. Repeat Example 5.6 for a triangular wave signal of period $N = 10$ whose first period is given by [0 1 2 3 4 5 4 3 2 1], where the first data point corresponds to $k = 0$.

DISCRETIZATION

Consider the sampled-data system in Figure 5.9, where the signal $e(t)$ is sampled at a period of T_s to obtain the sampled signal $e^*(t)$, which is the input to a ZOH yielding the signal $\bar{e}(t)$. The signal $\bar{e}(t)$ is the input to the continuous-time plant $G_p(s)$, whose output is $y(t)$. Because the output of the ZOH is a piecewise-constant signal, the output $y(t)$ consists of a series of step responses, known as *intersample ripples.* If we are interested only in the output $y(t)$ at the sampling instants, the ZOH and $G_p(s)$ can be combined to form a discrete equivalent, denoted by $G(z)$. This discrete equivalent, also known as the *pulse transfer function*, yields the same time response *at the sampling instants* as the continuous-time system.

The discrete equivalent $G(z)$ can be obtained using either the z-transform for transfer functions or the state-transition matrix for state-space models. Given $G_p(s)$ as a continuous-time transfer function, and using the ZOH transfer function as $G_{h0}(s) = (1 - \epsilon^{-sT_s})/s$, the discrete equivalent is obtained as

$$G(z) = \mathcal{Z}\{G_p(s)G_{h0}(s)\} \tag{5.4a}$$

$$= \mathcal{Z}\{1 - \epsilon^{-sT_s}\}\mathcal{Z}\left\{\frac{G_p(s)}{s}\right\} \tag{5.4b}$$

$$= (1 - z^{-1})\mathcal{Z}\left\{\frac{G_p(s)}{s}\right\} \tag{5.4c}$$

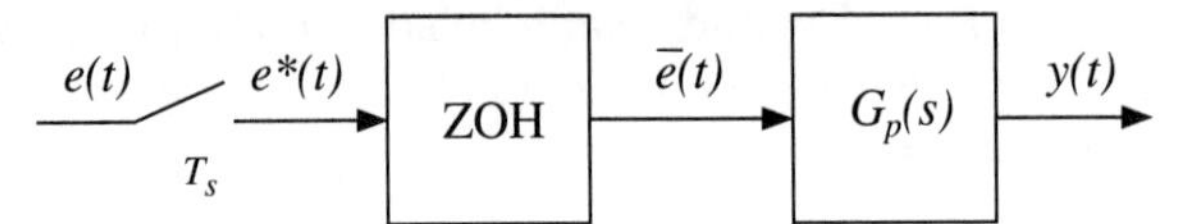

FIGURE 5.9 *An open-loop sampled-data control system*

To compute the z-transform $\mathcal{Z}\{G_p(s)/s\}$, we perform a partial-fraction expansion of $G_p(s)/s$ such that each individual first-order rational function of the Laplace variable s can be transformed into a corresponding first-order rational function of the discrete variable z according to

$$\mathcal{Z}\left\{\frac{1}{s+a}\right\} = \frac{1}{1-\epsilon^{-aT_s}z^{-1}} = \frac{z}{z-\epsilon^{-aT_s}}$$

where T_s is the sampling period.

If the continuous-time plant $G_p(s)$ is given in the state-space form

$$\dot{\mathbf{x}}(t) = \mathbf{A}\mathbf{x}(t) + \mathbf{B}\mathbf{u}(t), \quad \mathbf{y}(t) = \mathbf{C}\mathbf{x}(t) + \mathbf{D}\mathbf{u}(t) \tag{5.5}$$

the discrete equivalent $G(z)$ of $G_p(s)$ driven by a ZOH is

$$\mathbf{x}(k+1) = \mathbf{A}_d\mathbf{x}(k) + \mathbf{B}_d\mathbf{u}(k), \quad \mathbf{y}(k) = \mathbf{C}_d\mathbf{x}(k) + \mathbf{D}_d\mathbf{u}(k) \tag{5.6}$$

where

$$\mathbf{A}_d = \epsilon^{\mathbf{A}T_s}, \quad \mathbf{B}_d = \left[\int_0^{T_s} \epsilon^{\mathbf{A}(T_s-\tau)}d\tau\right]\mathbf{B}, \quad \mathbf{C}_d = \mathbf{C}, \quad \mathbf{D}_d = \mathbf{D}$$

and $\epsilon^{\mathbf{A}t}$ is the state transition matrix of the continuous-time system (5.5).

The Control System Toolbox provides the `c2d` function (continuous-time to discrete-time system conversion) with the `zoh` option to compute the discretization of the sampled-data system in Figure 5.9. The `c2d` function starts with any LTI continuous-time system object and returns the discrete equivalent of the same object type.

Another hold circuit that can be used is the first-order (triangular) hold (FOH), described in Exploratory Problem EP5.1. The discretization formula is more complex and will not be listed here. Interested readers can find it in advanced digital control texts. The FOH discretization can also be computed using `c2d` by specifying the method as `foh`.

☐ EXAMPLE 5.7 *ZOH Discretization*

Find the ZOH discretization of the continuous-time system represented by the state-space model

$$\mathbf{A} = \begin{bmatrix} -4 & -3 \\ 1 & 0 \end{bmatrix}, \quad \mathbf{B} = \begin{bmatrix} 1 \\ 0 \end{bmatrix}$$

$$\mathbf{C} = \begin{bmatrix} 1 & 5 \end{bmatrix}, \quad \mathbf{D} = \begin{bmatrix} 0 \end{bmatrix}$$

using a sampling period of $T_s = 0.1$ s.

Solution

In Script 5.7, the state matrices are entered to form the SS object. Then the function c2d is applied to find the discrete equivalents using a ZOH. From the MATLAB computation, the ZOH discrete-time system is

$$\mathbf{A}_d = \begin{bmatrix} 0.6588 & -0.2460 \\ 0.0820 & 0.9868 \end{bmatrix}, \quad \mathbf{B}_d = \begin{bmatrix} 0.0820 \\ 0.0044 \end{bmatrix}$$

with the $\mathbf{C}_d$ and $\mathbf{D}_d$ matrices the same as in the continuous-time system. Note that $\mathbf{A}_d$ can also be obtained from the MATLAB expm(A*Ts) command.[2] The order of the discrete-time model is the same as the continuous-time model. There is no increase in the number of states due to the ZOH.

MATLAB Script

```
% Script 5.7:  Discretization
A = [-4 -3; 1 0], B = [1; 0]       % state-space matrices
C = [1 5], D = 0
Gp = ss(A,B,C,D)                   % continuous-time state-space object
Ts = 0.1                           % sampling time
Gz_zoh = c2d(Gp,Ts,'zoh')          % zero-order hold discretization
```

□

WHAT IF? Redo Example 5.7 with the ZOH replaced by a first-order hold (FOH) and identify the similarities and differences of the two discrete-time models. ■

REINFORCEMENT PROBLEMS

In Problems 5.9 to 5.11, use the c2d function to find the discrete equivalent of the sampled-data system in Figure 5.9 with $G_p(s)$ and T_s given below, first using a ZOH and then using a FOH.

P5.9 Continuous-time TF model.

$$G_p(s) = \frac{2(s+2)}{s^2+4s+3} \quad \text{and} \quad T_s = 0.05 \text{ s}$$

P5.10 Continuous-time state-space model.

$$\mathbf{A} = \begin{bmatrix} -5 & -2 & 0 \\ 1 & 0 & 0 \\ 0 & 1 & 0 \end{bmatrix}, \quad \mathbf{B} = \begin{bmatrix} 1 \\ 0 \\ 0 \end{bmatrix}$$

$$\mathbf{C} = \begin{bmatrix} 1 & 0 & 2 \end{bmatrix}, \quad \mathbf{D} = \begin{bmatrix} 0 \end{bmatrix}, \quad \text{and} \quad T_s = 0.1 \text{ s}$$

[2] It is erroneous to use the exp(A*Ts) command, which gives a 2×2 matrix whose (i,j) entry is $\epsilon^{a_{ij}T_s}$, where a_{ij} is the (i,j) entry of $\mathbf{A}$.

P5.11 Continuous-time ZPK model.

$$G_p(s) = \frac{10(s+1)(s+10)}{(s+0.1)(s+5)(s+50)} \quad \text{and} \quad T_s = 0.01 \text{ s}$$

CLOSED-LOOP SAMPLED-DATA SYSTEMS

In the previous section, we examined the discretization of an open-loop continuous-time system driven by a zero-order hold (ZOH). Here, we study closed-loop sampled-data systems and compute the closed-loop pulse transfer function. In Figure 5.10, $G_p(s)$ denotes a continuous-time plant with the input $u(t)$ and the output $y(t)$. The output y is measured by the sensor whose transfer function is denoted by the continuous-time model $H(s)$. The measured signal is compared to the continuous-time command (reference) input $r(t)$ to generate the error signal $e(t)$, which is sampled, producing the impulse-sampled error signal $e^*(t)$. The input to the digital controller $G_c(z)$ is the sampled error e^*, and the output of $G_c(z)$ is used to drive a ZOH $G_{h0}(s)$ that provides a piecewise constant control signal to $G_p(s)$.

If we are interested only in the output $y(t)$ of the continuous-time plant $G_p(s)$ in Figure 5.10 *at the sampling instants*, we can use a closed-loop discrete equivalent system whose z-transform can be expressed as

$$T(z) = \frac{G(z)G_c(z)}{1 + \overline{HG}(z)G_c(z)} \tag{5.7}$$

where

$$G(z) = \mathcal{Z}\{G(s)\} = \mathcal{Z}\{G_p(s)G_{0h}(s)\},$$

and

$$\overline{HG}(z) = \mathcal{Z}\{H(s)G(s)\} = \mathcal{Z}\{\overline{HG}(s)\}$$

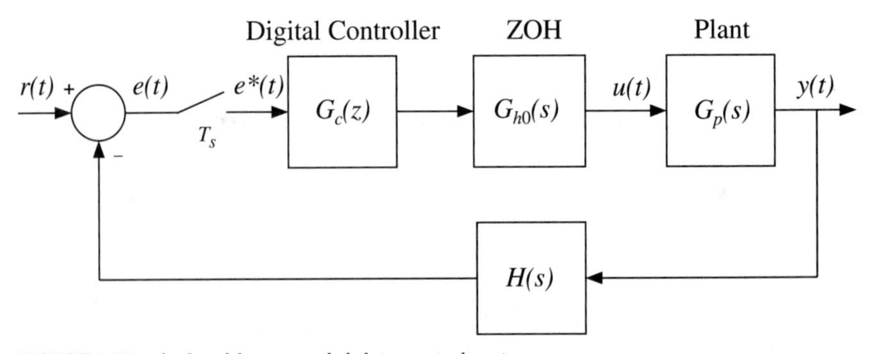

FIGURE 5.10 *A closed-loop sampled-data control system*

Note that $H(s)$ cannot be separated from $G_p(s)$ in the discretization because there is no sampler between the plant $G_p(s)$ and the sensor $H(s)$. This fact is denoted by the bar in the expressions $\overline{HG}(s)$ and $\overline{HG}(z)$. Also note that we write $G(z)G_c(z)$ for the forward-path transfer function, with $G(z)$ preceding $G_c(z)$, as was explained in the discussion of series connections in Chapter 3. If the sensor has unity DC gain and its dynamics are neglected, we can write $H(s) = 1$, in which case we have a unity feedback system and (5.7) reduces to

$$T(z) = \frac{G(z)G_c(z)}{1 + G(z)G_c(z)} \tag{5.8}$$

In Example 5.8, we will use MATLAB to compute the closed-loop pulse transfer functions for a unity feedback system and for a system for which $H(s) \neq 1$.

☐ EXAMPLE 5.8 *Closed-Loop Sampled-Data Control System*

Find the closed-loop system transfer function of the sampled-data control system in Figure 5.10 with

$$G_p(s) = \frac{1}{s(s+1)}, \quad G_c(z) = 5\left(\frac{z-0.9}{z-0.7}\right)$$

and $T_s = 0.1$ s, given

a. an ideal sensor, for which $H(s) = 1$

b. a slow sensor, described by $H(s) = 10/(s+10)$

Plot the step responses of the two discrete-time closed-loop systems on a single set of axes.

Solution

For both parts of the problem, we use the `c2d` command with the `zoh` option to set up $G(z)$ as `Gz` and to define $G_c(z)$ as `G_cz`. In part (a) of Script 5.8, we use the `feedback` function to compute the closed-loop transfer function according to (5.8). The result, denoted by `Ta`, is

$$T_a(z) = \frac{0.02419z^2 + 0.001626z - 0.02105}{z^3 - 2.581z^2 + 2.240z - 0.6544}$$

The poles of $T_a(z)$ are $z = 0.8960$ and $0.8546\epsilon^{\pm j0.1699}$, all of which are inside the unit circle, indicating that the closed-loop system is stable. Note that the second input to the `feedback` command, denoting the feedback path gain, is set to `1` because the gain in this path is unity. The step response of the closed-loop system with the ideal sensor is denoted by * in Figure 5.11. The steady-state value of the output is exactly unity because, for a constant input, the free integrator in the plant $G_p(s)$ forces the difference between the output and input signals to approach zero.

In part (b) of the script, we form the series connection $H(s)G_p(s)$ and use `c2d` with the `zoh` option to find $\overline{HG}(z)$, denoted as `HGz`. Because the

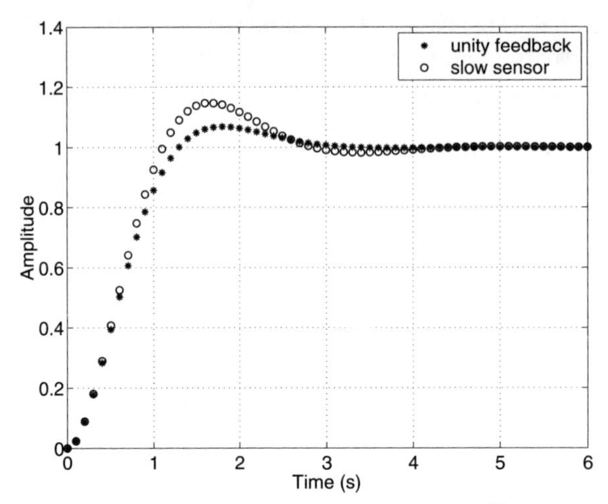

FIGURE 5.11 *Step responses using unity feedback (*) and slow sensor (o) for Example 5.8*

feedback command does not handle the closed-loop transfer function (5.7), we enter the formula in literal form as `Tb = Gz*G_cz/(1+HGz*G_cz)`. Because of the LTI system objects we have defined, MATLAB can interpret the expression correctly. Once the closed-loop transfer function $T_b(z)$ has been computed, however, its order has to be reduced because some of the poles of $G(z)G_c(z)$ and $\overline{HG}(z)G_c(z)$ can be cancelled. This cancellation is achieved by the MATLAB minimal realization function `minreal`, yielding the transfer function of the closed-loop system with the "slow" sensor as

$$T_b(z) = \frac{0.02419z^3 - 0.007272z^2 - 0.02165z + 0.007746}{z^4 - 2.966z^3 + 3.211z^2 - 1.471z + 0.2297}$$

The poles of $T_b(z)$ are $z = 0.8968, 0.8874\epsilon^{\pm j0.1856}$, and 0.3252, all of which are also inside the unit circle. Note that the last pole is present due to the sensor. The unreduced system is 7th order, so three pairs of coincidental poles and zeros were removed by the `minreal` command.

The step response of $T_b(z)$ is shown by the empty circles in Figure 5.11. Note that because the DC gain of $H(s)$ is unity, the steady-state output of $T_b(z)$ is also unity, resulting in a zero steady-state error between the input and output signals. Comparing the step responses of these two systems, we can see that the response with the sensor described by $H(s) = 10/(s + 10)$ has more overshoot because the sensor slows down the control action and hence reaches the unity steady-state condition after the unity-feedback response does.

MATLAB Script

```
% Script 5.8:  Step responses of closed-loop sampled-data systems
Gp = tf(1,[1 1 0])                   % plant in continuous tf form
fs = 10, Ts = 0.1                    % sampling frequency and period
Gz = c2d(Gp,Ts,'zoh')                % zero-order hold
G_cz = tf(5*[1 -0.9],[1 -0.7],Ts)    % controller in discrete tf form
%-------- Part a:  with ideal sensor --------------
```

cont.

```
Ta = feedback(Gz*G_cz,1,-1)          % CL transfer function
pole(Ta)                             % CL system poles
k  = [0:1:60]';
ya = step(Ta,k*Ts)                   % step response
%-------- Part b:  with slow sensor ---------------
H  = tf(10,[1 10])                   % sensor in continuous tf form
HGz = c2d(H*Gp,Ts,'zoh')             % discretize combined H & Gp
Tb = Gz*G_cz/(1+HGz*G_cz)            % CL transfer function
Tb = minreal(Tb)                     % eliminate duplicate states
pole(Tb)                             % CL system poles
yb = step(Tb,k*Ts)                   % step response
plot(k*Ts,ya,'*',k*Ts,yb,'o')        % plot step responses
legend('unity feedback','slow sensor')      % add legend
```

□

REINFORCEMENT PROBLEMS

Compute the discretized closed-loop sampled-data control system in Figure 5.10 for the plants and controllers given below. Find the closed-loop poles and the step response.

P5.12 ZPK model with $H(s) = 1$.

$$G_p(s) = \frac{10}{s(s+10)}, \quad G_c(z) = \frac{z-0.5}{z-0.8}, \quad \text{and} \quad T_s = 0.1 \text{ s}$$

P5.13 State-space model with $H(s) = 1$. The plant $G_p(s)$ is given by (5.5) with

$$\mathbf{A} = \begin{bmatrix} -5 & -2 & 0 \\ 1 & 0 & 0 \\ 0 & 1 & 0 \end{bmatrix}, \quad \mathbf{B} = \begin{bmatrix} 1 \\ 0 \\ 0 \end{bmatrix}$$

$$\mathbf{C} = \begin{bmatrix} 1 & 0 & 2 \end{bmatrix}, \quad \mathbf{D} = \begin{bmatrix} 0 \end{bmatrix}, \quad \text{and} \quad T_s = 0.05 \text{ s}$$

The controller $G_c(z)$ is given by

$$\mathbf{A}_K = 0.6, \quad \mathbf{B}_K = 1, \quad \mathbf{C}_K = 0.5, \quad \mathbf{D}_K = 0$$

P5.14 ZPK feedback system with $H(s) \neq 1$. Redo Problem 5.12 with $H(s) = 20/(s+20)$.

P5.15 State-space feedback system with $H(s) \neq 1$. Redo Problem 5.13 with $H(s) = 20/(s+20)$.

EXPLORATORY PROBLEMS

EP5.1 Interactive exploration. The ZOH functions similarly to a zero-order expansion of a Taylor series. A higher-order approximation requires taking into account the slope of the data points. In a first-order hold, the output is defined as

$$\bar{e}_1(t) = e(kT_s) + \alpha e'(kT_s)(t - kT_s), \quad kT_s \leq t < (k+1)T_s \tag{5.9}$$

where the slope $e'(kT_s)$ is given by

$$e'(kT_s) = \frac{e(kT_s) - e((k-1)T_s)}{T_s}$$

and $\alpha = 1$. Equation (5.9) also describes a fractional-order hold for $0 < \alpha < 1$. The RPI function `foh` can be used to generate the time-domain output of a fractional-order hold. Apply the `foh` function to the signals in the examples in this chapter and compare the results to the outputs of the ZOH.

COMPREHENSIVE PROBLEMS

CP5.1 Drumbeat sound track. A drum beat sound track was recorded by sampling the signal at $f_s = 16{,}384$ Hz and stored in the `local.mat` file. The sound track is contained in the `local` array. Use `fft` to find the spectrum of the signal. In particular, the sound track contains high-frequency noise components at about 7.1 kHz. You can hear these at the beginning and the end of the sound track when you play it using the MATLAB `sound` command. The signal can be downsampled to a lower frequency by skipping uniformly some of the data elements in `local`. For example, to sample at $f_s/2 = 8{,}192$ Hz, we skip every other element in `local`. Consider downsampling at $f_s/2$, $f_s/4$, and $f_s/8$ to observe whether any significant aliasing occurs, by analyzing the resulting frequency spectra. In addition, listen to the downsampled signals to hear how the tone of the high-frequency noise changes with the sampling frequency.

CP5.2 Frequency response of a first-order hold. Start from Equation (5.9) to derive the transfer function of the first-order hold. Then redo Example 5.5 to plot the frequency response of the first-order hold and Example 5.6 to investigate the Gibbs phenomenon using only the five lowest frequency components of the first-order hold output.

CP5.3 Anti-aliasing filters. To counter any aliasing effect due to sampling, a continuous-time input signal often undergoes lowpass filtering before sampling to remove any unwanted high-frequency components. To illustrate this process with MATLAB, consider the signal

$$e(t) = 5\cos(4\pi t) + 2\cos(18\pi t)$$

with frequency components at 2 and 9 Hz. If the signal is to be sampled directly at 10 Hz, a 1 Hz component will appear. However, if $e(t)$ is passed into an analog lowpass filter

$$H(s) = \left(\frac{10\pi}{s + 10\pi}\right)^n$$

having a corner frequency at 10π rad/s = 5 Hz, where n is the number of stages of the filter, the magnitude of the 9 Hz component will be reduced. Find n such that the 1 Hz component of the sampled signal is less than 5% of the amplitude of the 2 Hz component. Demonstrate the design in MATLAB by using time increments of 0.001 sec to simulate the analog filtering process. Then sample the output of the signal at 10 Hz to obtain the discrete-time signal. Finally, apply the `fft` function to find the spectrum of the sampled signal.

CP5.4 Ball and beam system. Run the file `dbbeam.m` to obtain the continuous-time state-space model from the input voltage to the wheel angle θ and the ball position ξ. Discretize the continuous-time model at a sampling frequency of 50 Hz and compare the result to the discrete-time model computed by `dbbeam.m`. Examine the structure of the discretized A_d and B_d matrices and compare the location of the zero and nonzero entries of these matrices with their continuous counterparts. Compute the poles and zeros for each output of the discretized system. Establish the correspondence between the continuous-time and discrete-time model poles and zeros.

Repeat the above activities by sampling the system first at 25 Hz and then at 100 Hz. How do the poles and zeros of the resulting discretized systems vary from those in the system sampled at 50 Hz?

CP5.5 Electric power system. Run the file `dpower.m` to obtain the continuous-time state-space model from the input (V_{field}) to each of the three outputs (V_{term}, ω, and P). Use the `c2d` command to discretize the continuous-time model at a sampling frequency of 100 Hz and compare the result to the discrete-time model computed by `dpower.m`. Examine the structure of the discretized $\mathbf{A}_d$ and $\mathbf{B}_d$ matrices and compare the locations of the zero and nonzero entries of these matrices with their continuous counterparts. Compute the poles and zeros for each output of the discretized system. Establish the correspondence between the continuous-time and discrete-time model poles. In particular, note that the complex continuous-time poles become complex discrete-time poles.

The continuous-time model has a pair of complex poles at $s = -0.479 \pm j9.33$ whose frequency is $9.33/2\pi = 1.48$ Hz. Discretize the system at 2 Hz and examine the resulting complex poles. Compare the oscillations in the electrical power variable of the open-loop system discretized at 100 Hz and at 2 Hz to a 0.05 step in V_{ref}.

CP5.6 Hydroturbine system. Run the file `dhydro.m` to obtain the continuous-time state-space model from the input u to the output power P of the hydroturbine system. Discretize the continuous-time model at a sampling frequency of 10 Hz and compare the result to the discrete-time model computed by `dhydro.m`. Compute the poles and zero of the discretized system and establish their correspondence to the continuous-time model poles and zero. Confirm the fact that the right-half-plane zero of the continuous-time plant has been converted to a zero outside the unit circle in the discretized model.

SUMMARY

In this chapter we have examined the sampling of continuous-time signals to obtain discrete-time signals as well as the reconstruction of discrete-time signals to obtain continuous-time signals. This process allowed us to derive equivalent discrete-time systems for sampled-data control systems. Aliasing was illustrated and discussed in both the time and the frequency domain. The zero-order hold (ZOH) was analyzed and applied to the discretization of sampled-data control systems in transfer-function form and state-space form. Following the discretization, closed-loop sampled-data control systems using a digital controller and including sensor dynamics were derived and illustrated with an example.

ANSWERS

P5.4 Aliasing occurs when sampling at 30 Hz. No aliasing occurs at 40 Hz.

P5.5 One such signal is $\sin(200 \times 2\pi t + \pi/2)$.

P5.6 $e_2^*(kT_s) = (-1)^k \delta(t - kT_s)$

P5.9 Zero-order hold: $G(z) = (0.0952z - 0.08614)/(z^2 - 1.812z - 0.8187)$. First-order hold: $G(z) = (0.04838z^2 + 0.003019z - 0.04234)/(z^2 - 1.812z - 0.8187)$.

P5.10 Zero-order hold:

$$\mathbf{A}_d = \begin{bmatrix} 0.5993 & -0.1569 & 0 \\ 0.07843 & 0.9914 & 0 \\ 0.004254 & 0.09971 & 1 \end{bmatrix}, \quad \mathbf{B}_d = \begin{bmatrix} 0.07843 \\ 0.004254 \\ 0.0001476 \end{bmatrix}$$

$\mathbf{C}_d = \mathbf{C}$ and $\mathbf{D}_d = \mathbf{D}$.

First-order hold:

$$\mathbf{B}_d = \begin{bmatrix} 0.06116 \\ 0.007579 \\ 0.0004758 \end{bmatrix}, \quad \mathbf{D}_d = \begin{bmatrix} 0.04262 \end{bmatrix}$$

$\mathbf{A}_d$ and $\mathbf{C}_d$ matrices are the same as those obtained with the zero-order hold.

P5.11 Zero-order hold:

$$G(z) = \frac{0.081178(z-0.905)(z-0.9901)}{(z-0.6065)(z-0.9512)(z-0.999)}$$

First-order hold:

$$G(z) = \frac{0.043476(z-0.99)(z-0.9048)(z+0.8636)}{(z-0.6065)(z-0.9512)(z-0.999)}$$

P5.12 Closed-loop transfer function:

$$T(z) = \frac{0.03679z^2 + 0.00803z - 0.01321}{z^3 - 2.131z^2 + 1.47z - 0.3075}$$

Closed-loop poles are $z = 0.8925\epsilon^{\pm j0.2119}$ and 0.3861.

P5.13 The state-space matrices of the closed-loop discrete-time system are

$$\mathbf{A}_d = \begin{bmatrix} 0.7767 & -0.08841 & 0 & 0.02210 \\ 0.04420 & 0.9977 & 0 & 0.0005758 \\ 0.001152 & 0.04996 & 1 & 0 \\ -1 & 0 & -2 & 0.6 \end{bmatrix}, \quad \mathbf{B}_d = \begin{bmatrix} 0 \\ 0 \\ 0 \\ 1 \end{bmatrix}$$

$$\mathbf{C}_d = \begin{bmatrix} 1 & 0 & 2 & 0 \end{bmatrix}, \quad \mathbf{D}_d = \begin{bmatrix} 0 \end{bmatrix}$$

Closed-loop poles are $z = 0.7033\epsilon^{\pm j0.1715}$ and $0.9947\epsilon^{\pm j0.03154}$.

P5.14 Closed-loop system

$$T(z) = \frac{0.03679z^3 + 0.003051z^2 - 0.0143z + 0.001788}{z^4 - 2.286z^3 + 1.781z^2 - 0.5055z + 0.03794}$$

Closed-loop poles are $z = 0.9071\epsilon^{\pm j0.2177}$, 0.3996, and 0.1154.

P5.15 The state-space matrices of the closed-loop discrete-time system are

$$\mathbf{A}_d = \begin{bmatrix} 1.343 & -0.4739 & 0.2922 & -0.1013 & 0.1583 \\ 0.5321 & 0.4433 & 0.3931 & -0.09525 & -0.04902 \\ 0.1246 & 0.1180 & 0.8940 & -0.01898 & 0.1070 \\ 4.708 & -4.156 & 2.939 & 0.08083 & 0.5595 \\ -0.2702 & 0.2483 & -0.1244 & 0.04421 & 0.9809 \end{bmatrix}, \quad \mathbf{B}_d = \begin{bmatrix} 0.2171 \\ 0.09713 \\ 0.04929 \\ 1.392 \\ -0.06755 \end{bmatrix}$$

$$\mathbf{C}_d = \begin{bmatrix} -0.2374 & -0.1691 & 0.4179 & -0.03860 & -1.496 \end{bmatrix}, \quad \mathbf{D}_d = \begin{bmatrix} 0 \end{bmatrix}$$

Closed-loop poles are $z = 0.3084, 0.7380\epsilon^{\pm j0.2066}$, and $0.9951\epsilon^{\pm j0.03148}$.

6 *Frequency Response, Digital Filters, and Discrete Equivalents*

PREVIEW

The frequency response of a discrete-time system has many uses. It can be used to predict the output of a system given an input signal from a frequency-domain viewpoint. Such knowledge is important in designing digital filters, which are useful in digital signal processing. Frequency response can also be used to determine feedback system stability. In this chapter we show the computation of frequency response and its use in determining sinusoidal steady-state response. As a direct application of frequency response, we discuss simple digital filters, whose counterparts in control are lead-lag compensators. Finally, we present sampled-data transformations, which are useful for transforming analog filters and controllers into digital equivalents.

FREQUENCY RESPONSE

The frequency response of a discrete-time system $G(z)$ is defined as

$$G(z)|_{z=\epsilon^{j\omega T_s}} = G(\epsilon^{j\omega T_s}), \quad 0 \leq \omega \leq \pi/T_s \tag{6.1}$$

where T_s is the sampling time. In words, $G(z)$ is evaluated for z on the circumference of the upper half of the unit circle in the z-plane. The frequency response $G(\epsilon^{j\omega T_s})$ is a complex function; it is usually plotted with its magnitude versus frequency and its phase versus frequency. Note that on the lower half of the unit circle, the frequency response is the complex conjugate of that on the upper half—that is,

$$G(\epsilon^{-j\omega T_s}) = G(\epsilon^{j\omega T_s})^* \quad \text{where} \quad 0 \leq \omega \leq \pi/T_s$$

Thus, computing (6.1) is sufficient to generate the complete frequency response. Also note that at $\omega = 0$, (6.1) yields the DC gain, and $\omega = \pi/T_s$ corresponds to the frequency $\omega_s/2$, where $\omega_s = 2\pi/T_s$ is the sampling frequency. In contrast to the frequency response of a continuous-time system, the frequency response of a discrete-time system can be completely determined in a *finite* frequency range from 0 to $\omega_s/2$. MATLAB provides the `bode` function to compute the frequency response for both continuous-time and discrete-time systems. When the `bode` command is applied to a discrete-time LTI object, (6.1) is used in computing the frequency response.

Example 6.1 illustrates the use of `bode` to find the frequency response of a discrete-time system. The example also illustrates the use of the *data marker* capability of the LTI plotting functions in the Control System Toolbox to identify precise values on the response plots.

☐ EXAMPLE 6.1 *Lowpass Butterworth Filter*

A second-order lowpass Butterworth filter with a cutoff frequency of $\omega = 0.4\pi/T_s$ rad/s is

$$G(z) = \frac{0.2066(z^2 + 2z + 1)}{z^2 - 0.3695z + 0.1958} \tag{6.2}$$

where the sampling period is $T_s = 0.1$ s. Plot the poles, the zeros, and the frequency response of $G(z)$. Find the gain of the filter as the magnitude ratio and in dB at $\omega = 0, 0.4\pi/T_s$, and π/T_s rad/s.

Solution

After the discrete-time LTI object `G` has been created, the `pzmap` function is used to plot the poles and zeros. As shown in Figure 6.1(a), the poles are at $z = 0.1848 \pm j0.4021$, and there are two zeros at $z = -1$. When the `bode` function is applied with `G` as the input argument, the frequency-response plot shown in Figure 6.1(b) is generated. Note the lowpass characteristic of the filter. To find the frequency response at specific frequencies, we use the `bode` command again, with the frequency vector `ww` as its second input argument, and `mag_ratio` and `ph` as the output arguments.

At $\omega = 0$—that is, at DC—the gain is unity (0 dB). At the cutoff frequency $\omega = 0.4\pi/T_s$, the gain is 0.7071 (−3 dB). At half the sampling frequency ($\omega =$

π/T_s), the gain is identically zero ($-\infty$ in dB, which MATLAB represents as `-Inf`), due to the zeros at $z = -1$. The commands are given in Script 6.1.

In the frequency response plot of Figure 6.1(b), we can click the left mouse button on the response curve where the magnitude is close to -3 dB. Then the data marker function of the Control System Toolbox displays the magnitude and frequency of the selected point, as shown in the upper plot of Figure 6.1(b). Furthermore, the selected data point can be moved by holding down the left button and dragging the pointer along the response curve until the precise data point is found. A right click on the data marker box displays a menu of options including the placement of the data marker box. We can accurately find the phase of the frequency response at the -3 dB frequency in a similar manner, as shown in the lower plot of Figure 6.1(b).

MATLAB Script

```
% Script 6.1:  Frequency Response Plot
Ts = 0.1                               % sampling period
num = 0.2066*[1 2 1]                   % filter numerator
den = [1 -0.3695 0.1958]               % filter denominator
G = tf(num,den,Ts)                     % LTI object in TF form
ucircle,axis equal,hold on
pzmap(G),hold off                      % pole-zero plot
w = logspace(-2,0,100)*pi/Ts;          % freq vector ending at w = pi*Ts
bode(G,w)                              % make frequency response plot
%--  compute magnitudes at 3 specified frequencies ---
ww = [0 0.4*pi/Ts pi/Ts]
[mag_ratio,ph] = bode(G,ww)
mag_db = 20*log10(mag_ratio)
disp(['at frequencies ',vec2str(ww)])
disp('filter gains are ')
disp([vec2str(mag_ratio),' magnitude ratio'])
disp([vec2str(mag_db),' dB'])
```

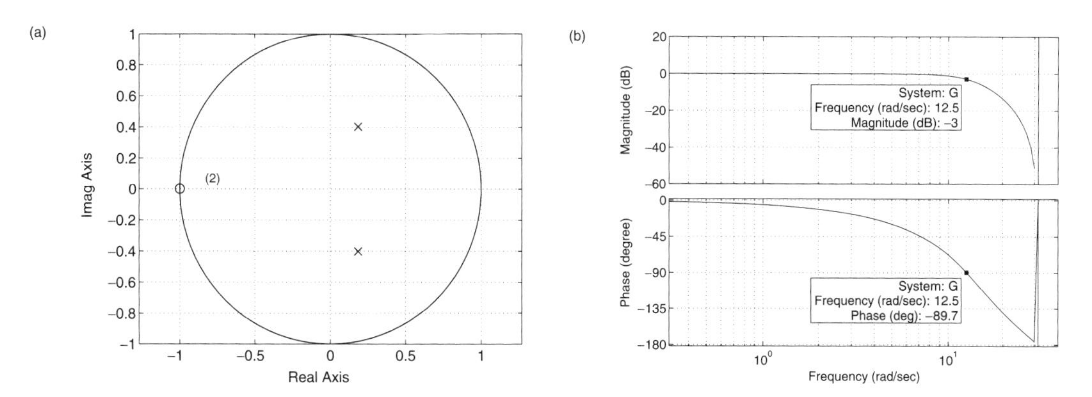

FIGURE 6.1 *Lowpass Butterworth filter for Example 6.1: (a) pole-zero plot; (b) frequency-response plot*

□

REINFORCEMENT PROBLEMS

For the digital filters listed below, plot the poles and zeros and the frequency response using a sampling period of $T_s = 0.01$ s. Use the magnitude plot to justify the filter's designation as highpass, bandpass, or bandstop.[1]

P6.1 Second-order Butterworth highpass filter.

$$G(z) = \frac{0.2066(z^2 - 2z + 1)}{z^2 + 0.3695z + 0.1958}$$

P6.2 Fourth-order Chebyshev I bandpass filter.

$$G(z) = \frac{0.0931(z^4 - 2z^2 + 1)}{z^4 + 1.0349z^2 + 0.4293}$$

P6.3 Fourth-order Chebyshev II bandstop filter.

$$G(z) = \frac{0.4207(z^4 + 1.7994z^2 + 1)}{z^4 + 0.3773z^2 + 0.2210}$$

SINUSOIDAL STEADY-STATE RESPONSE

One of the many uses of the frequency response is the computation of the steady-state response of a stable LTI discrete-time system to a sinusoidal input signal. Given a system $G(z)$ with an input $e(kT_s) = A\cos(k\omega_0 T_s + \theta)$ and an output $y(kT_s)$, the steady-state response of the system is

$$y_{\text{ss}}(kT_s) = AM\cos(k\omega_0 T_s + \theta + \phi) \tag{6.3}$$

where M and ϕ are the magnitude and phase of the frequency response of $G(z)$ evaluated at $\omega = \omega_0$—that is,

$$G(z)|_{z=\epsilon^{j\omega_0 T_s}} = G(\epsilon^{j\omega_0 T_s}) = M\epsilon^{j\phi} \tag{6.4}$$

The results given by (6.3) and (6.4) can be derived using the z-transform and partial-fraction expansion. We demonstrate these relationships in Example 6.2.

☐ EXAMPLE 6.2 *Sinusoidal Steady-State Response*

Let the input signal to the lowpass Butterworth filter used in Example 6.1 be

$$e(kT_s) = 5\cos(32\pi kT_s + 30°)$$

Find the steady-state output y_{ss} and compare it to the steady-state response using (6.3). Assume a sampling period of $T_s = 0.01$ s.

[1]The characteristics of these filters are discussed in a later section on digital filters.

Solution

In Script 6.2, we define the input signal `e` and the discrete-time LTI object `G` and then use `lsim` to generate the system response. The magnitude and phase of the frequency response at the input frequency $\omega_0 = 32\pi$ rad/s are computed by the `bode` command to be 0.8679 and $-68.22°$, respectively. These values are then used in (6.3) to compute the steady-state response. Figure 6.2 shows the input signal, the complete response, and the steady-state response. We have connected the points with lines to make it easier to distinguish the signals. Keep in mind, however, that these are discrete-time signals that are undefined between the sample points. Figure 6.2 also illustrates the use of the `legend` command for identifying the individual plots.

Note that the complete response and the steady-state response are almost indistinguishable after 0.04 s. The amplitude of the steady-state response is 0.8679 times that of the input signal. The steady-state response is delayed from the input signal by 68.22°, which is equivalent to $(68.22/360) \times 1/16 = 0.0118$ s, where 1/16 s is the period of the input sinusoidal signal.

MATLAB Script

```
% Script 6.2:  Sinusoidal steady-state response
Ts = 0.01                              % sampling time
k = [0:1:20]'                          % generate time window
kTs = k*Ts
e = 5*cos(32*pi*kTs + 30*pi/180);      % input signal
num = 0.2066*[1 2 1]                   % Butterworth filter numerator
den = [1 -0.3695 0.1958]               % Butterworth filter denominator
G = tf(num,den,Ts)                     % LTI object
y = lsim(G,e)                          % simulate time response
[mag,ph] = bode(G,32*pi)               % mag and phase at w0
yss = 5*mag*cos(32*pi*kTs + (30+ph)*pi/180);     % eqn 6.3
plot(kTs,e,'o:',kTs,y,'*:',kTs,yss,'v:')  % plot all three signals
axis([0 0.2 -6 8]), grid               % change axis to allow space for legend
legend('input e(k*Ts)','filter output','steady-state output')
hold on
plot(kTs,0*kTs), hold off              % show horizontal axis as solid line
```

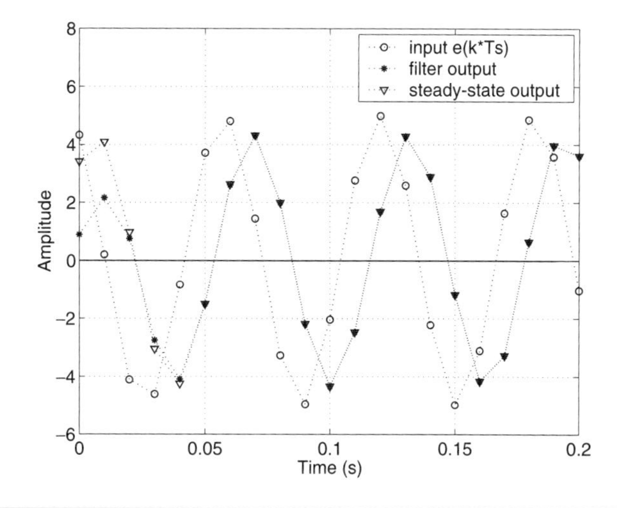

FIGURE 6.2 *Input signal, complete response, and steady-state response for Example 6.2*

WHAT IF? Find the steady-state output response when the frequency of the input signal in Example 6.2 is $\omega_0 = 50\pi$ rad/s. Use the plots of the magnitude and phase in Figure 6.1(b) to justify the changes in the steady-state magnitude and phase. ■

□

REINFORCEMENT PROBLEMS

P6.4 Highpass filter. The following signals are inputs to the highpass filter defined in Problem 6.1, using a sampling time $T_s = 0.01$ s. In each case, plot the magnitude and phase versus frequency and use (6.3) to verify the steady-state response. Compare the amplitudes of the outputs and explain why they differ.

(a) $e_1(kT_s) = 2\sin(4\pi kT_s + 20°)$

(b) $e_2(kT_s) = 2\cos(60\pi kT_s - 145°)$

P6.5 Chebyshev I bandpass filter. Repeat Problem P6.4 using the Chebyshev I filter from Problem 6.2.

P6.6 Chebyshev II bandstop filter. Repeat Problem P6.4 using the Chebyshev II filter from Problem 6.3.

DIGITAL FILTERS

The discussion in the previous section shows that with an appropriate frequency response, a discrete-time system can be used as a filter to modify the frequency content of the input signal. There are four different types of filters: lowpass, highpass, bandpass, and bandstop. In this section, we will study only first- and second-order filters. More detailed treatments of digital filters can be found in digital signal processing texts, such as Oppenheim and Shafer (1999). In Chapter 9, some of these simple filters will be used as lead or lag compensators obtained from frequency-response methods.

Consider the first-order digital filter whose transfer function is

$$G(z) = K\left(\frac{z - z_0}{z - z_p}\right)$$

where K is the gain, and the pole at $z = z_p$ and the zero at $z = z_0$ are both real and inside the unit circle. The DC gain of the filter can be computed by setting $z = 1$ in $G(z)$, yielding $K_{\text{DC}} = K(1 - z_0)/(1 - z_p)$. The high-frequency gain is evaluated at $\omega = \pi/T_s$, which corresponds to $z = -1$, resulting in $K_{\text{HF}} = K(1 + z_0)/(1 + z_p)$. Thus, $G(z)$ is a lowpass filter if $K_{\text{DC}} > K_{\text{HF}}$, which is the case when z_p is closer to the point $z = 1$ than z_0 is. On the other hand, $G(z)$ is a highpass filter if $K_{\text{DC}} < K_{\text{HF}}$, which is the case when z_0 is closer to $z = 1$ than z_p is. Example 6.3 illustrates the lowpass filtering of a multiple-frequency signal.

☐ EXAMPLE 6.3 *Response of a Lowpass Digital Filter*

The continuous-time periodic signal $e(t)$ defined by

$$e(t) = \cos(14\pi t) + \cos(50\pi t) + \cos(86\pi t) \tag{6.5}$$

has sinusoidal components at 7, 25, and 43 Hz. Sample the signal at 100 Hz to obtain $e(kT_s)$. Apply this signal to the first-order lowpass filter with transfer function

$$G(z) = 0.3250\left(\frac{z + 0.5385}{z - 0.5}\right)$$

and simulate the response for 10 s. Find the sinusoidal components in the last 1 s of the output signal ($9 \leq t \leq 10$ s) and comment on the effectiveness of the filter.

Solution

The zero of the filter transfer function is at $z = -0.5385$ and the pole is at $z = 0.5$. Hence, the pole is closer to $z = 1$ than the zero is. The gain $K = 0.3250$ results in unity DC gain. The lowpass characteristic of the filter is shown in Figure 6.3(a), with a high-frequency gain reduction of 20 dB. The bandwidth of the filter is that frequency for which the gain is 3 dB below the DC gain, which is approximately 10 Hz in this case.

From (6.5), we see that the input signal $e(kT_s)$, which is shown in the upper plot of Figure 6.3(b), has equal components at 7, 25, and 43 Hz. The output signal, shown in the lower part of the figure, has reduced high-frequency components relative to the input, as expected. To ensure that all the transients have decayed, and to be able to see the signal's values clearly, we display the data from the last 0.5 s of the time response only, in the lower plot of Figure 6.3(b).

Finally, we use the fast Fourier transform function `fft` to compute the spectral content of the output signal, shown in Figure 6.3(c). The magnitude (upper) plot indicates that the component at 43 Hz has been reduced to about 1/10 of its original magnitude and the component at 25 Hz is only about 30% of its original magnitude.

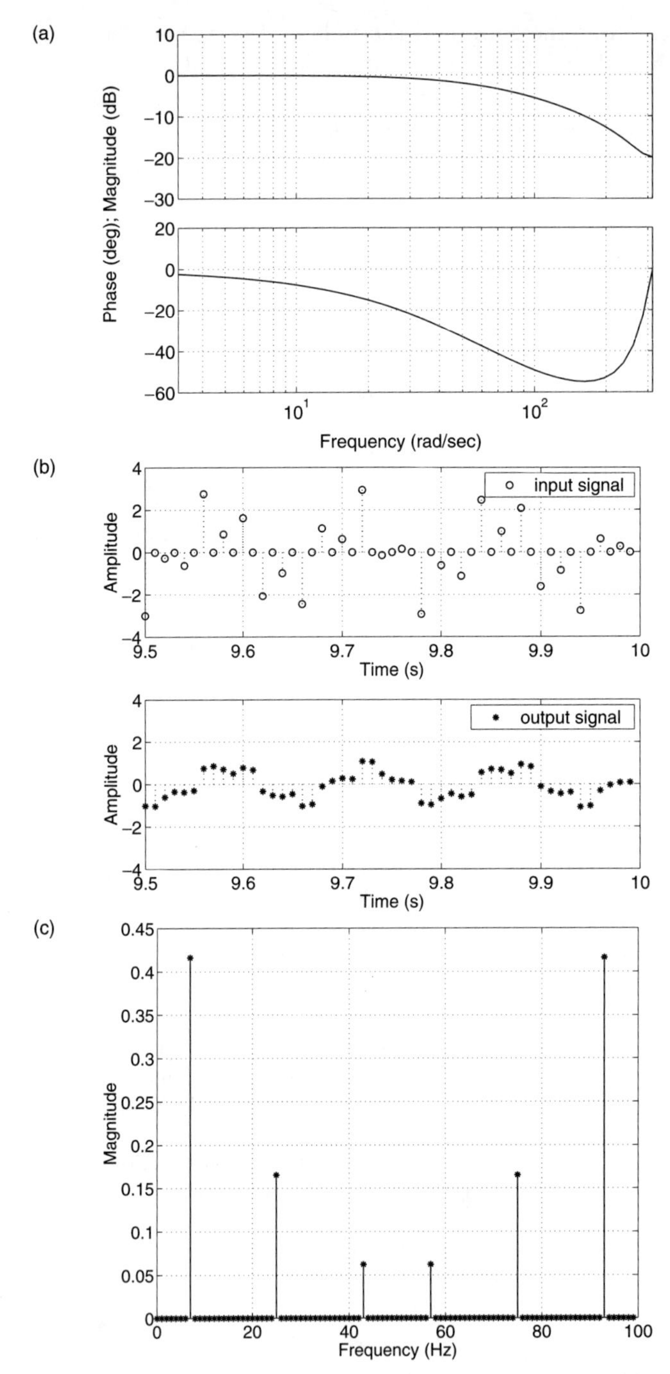

FIGURE 6.3 *Example 6.3: (a) frequency response of $G(z)$; (b) input and output signals; (c) frequency coefficient magnitudes of output signal*

MATLAB Script

```
% Script 6.3:  Lowpass filtering
Ts = 0.01                              % sampling period
G = tf(0.325*[1 0.5385],[1 -0.5],Ts) % create lowpass filter
w = logspace(-2,0,50)*pi/Ts;           % specify frequency range
bode(G,w)                              % frequency response plot
%
t = [0:Ts:10.0-Ts];                    % time array from 0 to 9.99 sec
e = cos(14*pi*t)+cos(50*pi*t)+cos(86*pi*t);  % DT signal
y = lsim(G,e);                         % filter output
t9 = [901:1000]';    % vector of time points during last second
%---- time plots -------
subplot(2,1,1)                         % split graphics window
stem(t(t9),e(t9),':'), grid
axis([9.5 10.0 -4 4])                  % set axis to accommodate legend
legend('input signal')
subplot(2,1,2)
stem(t(t9),y(t9),':*'), grid
axis([9.5 10.0 -4 4])                  % make axis same as input plot
legend('output signal')
%
N = length(y(t9))                      % number of data points
Yp = fft(y(t9))/N                      % compute spectral contents
ff = [0:1/Ts/N:(1.0-Ts)/Ts];           % frequency array
%- plot magnitude of spectrum vs frequency
stem(ff,abs(Yp),'*')
```

□

For bandpass and bandstop characteristics, we have to use filters of order two or greater. Consider the second-order filter whose transfer function is

$$G(z) = K\left[\frac{(z-\alpha\epsilon^{j\omega_0 T_s})(z-\alpha\epsilon^{-j\omega_0 T_s})}{(z-\beta\epsilon^{j\omega_0 T_s})(z-\beta\epsilon^{-j\omega_0 T_s})}\right] \tag{6.6}$$

where K is the gain and $0 \le \alpha, \beta < 1$. The poles and zeros are on the same rays from the origin $z = 0$. For $\alpha > \beta$, the zeros are closer to the unit circle than the poles are. Thus, the frequency response $|G(\epsilon^{j\omega T_s})|$ reaches its minimum at $\omega = \omega_0$, achieving a bandstop effect. On the other hand, for $\alpha < \beta$, the poles are closer to the unit circle than the zeros are. Thus, the frequency response $|G(\epsilon^{j\omega T_s})|$ reaches its maximum at $\omega = \omega_0$, achieving a bandpass effect. Example 6.4 illustrates bandpass filtering.

☐ EXAMPLE 6.4 *Bandpass Filtering*

Plot the poles and zeros and the frequency response of a filter that has the form of (6.6), with $T_s = 0.01$ s, $\omega_0 = 50\pi$ rad/s, $\alpha = 0.5$, $\beta = 0.95$, and $K = 0.13$. Apply the sampled signal $e(kT_s)$ from Example 6.3 to the filter and simulate the time response for 10 s. Plot the last 0.5 s of the output signal.

Solution

Following Script 6.4, we make the pole-zero plot in Figure 6.4(a), where both the poles and the zeros are purely imaginary. Because the poles are closer to the unit circle than the zeros are, the filter will exhibit a bandpass characteristic, which is verified by the frequency-response plot in Figure 6.4(b). The center frequency is $\omega_0 = 50\pi$ rad/s (25 Hz), with low- and high-frequency attenuations at about 20 dB. When the signal $e(kT_s)$ is passed through the filter, we expect only the 25 Hz component to pass through the filter and the other two frequencies (14 and 43 Hz) to be reduced by approximately 20 dB, which corresponds to a factor of 10. Figure 6.4(c) shows the output signal, which consists mainly of the 25 Hz component. This can easily be verified using the `fft` command to compute the magnitudes of the output's frequency coefficients, as was done in Example 6.3.

MATLAB Script

```
% Script 6.4:  Bandpass filtering
Ts = 0.01                              % sampling period
w_0 = 50*pi                            % center frequency
a = 0.5, b = 0.95, K = 0.13
%-- built bandpass filter, where j is sqrt(-1)
num = K*conv([1 -a*exp(j*w0*Ts)],[1 -a*exp(-j*w0*Ts)])
den = conv([1 -b*exp(j*w0*Ts)],[1 -b*exp(-j*w0*Ts)])
G = tf(num,den,Ts)                     % bandpass filter as TF object
ucircle,axis equal,hold on
pzmap(G),hold off                      % plot poles and zeros
%---- plot frequency response magnitude and phase ----
w = logspace(-2,0,200)*pi/Ts;          % specify frequency range
bode(G,w)                              % frequency response plot
%---- plot output of the filter ---
t = [0:Ts:10.0-Ts];                    % time array from 0 to 9.99 s
e = cos(14*pi*t)+cos(50*pi*t)+cos(86*pi*t);  % DT signal
y = lsim(G,e);                         % filter output for 0 --> 9.99 s
t95 = [951:1000]';                     % define time vector for last 0.5 s
stem(t(t95),y(t95),':o')
```

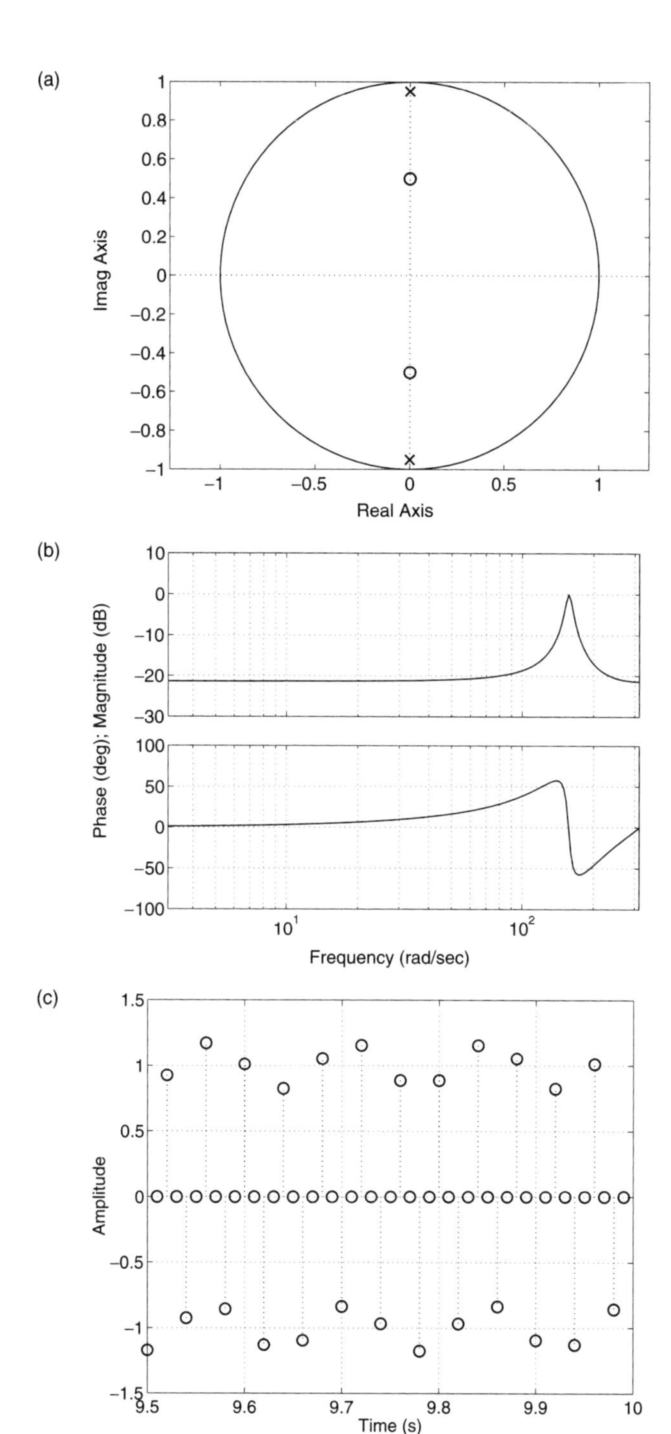

FIGURE 6.4 *Example 6.4: (a) pole-zero plot of bandpass filter; (b) frequency response of bandpass filter; (c) output signal of bandpass filter for* $9.5 < t < 10.0$ *s*

WHAT IF? Redo Example 6.4 with β reduced from 0.95 to 0.90 and investigate the effect on the filter's bandpass characteristics. Is the bandpass filtering as effective as for $\beta = 0.95$? ■

□

REINFORCEMENT PROBLEMS

P6.7 Highpass filter. Repeat Example 6.3 for the digital filter whose transfer function is

$$G(z) = 3.077\left(\frac{z - 0.5}{z + 0.5385}\right)$$

which is obtained by inverting the lowpass filter in the example.

P6.8 Bandstop filter. Repeat Example 6.4 with $\alpha = 0.95$, $\beta = 0.5$, and $K = 7.692$—the inverse of the filter in the example.

P6.9 Two filters in series. Connect the filters in Problems P6.7 and P6.8 in series and repeat Example 6.3.

DISCRETE EQUIVALENTS

One method of designing digital filters is to first obtain a satisfactory analog filter $G(s)$ and then derive from it a discrete equivalent filter $G(z)$ using one of several sampled-data transformations. Known as *indirect design*, this approach is advantageous in that continuous-time filters satisfying standard design specifications are readily available. Some common analog-to-digital filter conversion methods are:

1. *Impulse-invariant transformation.* The discrete equivalent filter $G(z)$ is obtained from the standard z-transform $G(z) = \mathcal{Z}\{G(s)\}$, such that the outputs of $G(s)$ and $G(z)$ subject to a unit-impulse input are identical at the sampling instants. A factor of T_s is applied to $G(z)$ to account for the change in gain by sampling.

2. *Bilinear (Tustin) transformation.* $G(z)$ is obtained by rewriting $G(s)$ with the complex variable s replaced by $2(z-1)/(T_s(z+1))$.

3. *Bilinear transformation with frequency prewarping.* The transformation first computes the scaling factor

$$a = \frac{2}{T_s}\tan\frac{\omega_c T_s}{2}$$

where ω_c is the critical frequency at which the characteristic of the discrete-time filter is identical to that of the continuous-time filter. Then the bilinear transformation is applied to the prewarped transfer function $G(s/a)$.

4. *Matched-z transformation.* All the s-plane poles and finite zeros of $G(s)$ are mapped to the poles and zeros of $G(z)$ according to $z = e^{sT_s}$. All but one of the infinite zeros of $G(s)$ are mapped to zeros at $z = -1$ of $G(z)$. The DC gain of $G(z)$ is selected to match the DC gain of $G(s)$. MATLAB provides the `c2d` function (continuous-to-discrete system conversion) to obtain discrete equivalents using the Tustin transformation and the matched-z transformation. The `c2d` function was used in Chapter 5 to discretize a sampled-data control system using the `'zoh'` and `'foh'` options. The impulse-invariant transformation can be computed using the RPI function `imp_inv`.

For digital control design in which an analog controller is first obtained, the bilinear transformation can be applied to the analog controller to obtain a discrete equivalent. This design method will be discussed in Chapter 9. The bilinear transformation is used because it preserves stability.

One way to evaluate a discrete equivalent $G(z)$ is to compare its frequency response with that of the continuous-time system $G(s)$. The frequency response for a continuous-time system is obtained by plotting the transfer function $G(s)$ along the $j\omega$-axis. For a discrete-time system, the frequency response is obtained by plotting $G(z)$ on the circumference of the unit circle, where $z = e^{j\omega T_s}$. Because of sampling, we plot only from $\omega = 0$ to $\omega = \pi/T_s$. Furthermore, to plot several frequency responses together, we need to use the `bode` function to generate the magnitude and phase arrays. The `bode` function outputs the frequency responses in cell arrays, however. We have coded an RPI function, `bodedb`, which gives the magnitude (in dB) and phase (in degrees) output variables directly as matrix arrays.

☐ EXAMPLE 6.5 *Discrete Equivalent of a Bandpass Filter*

A second-order analog Butterworth filter has the s-domain transfer function

$$G(s) = \frac{1}{s^2 + \sqrt{2}\, s + 1}$$

Find its discrete equivalents $G(z)$ using the methods listed in this section, assuming a sampling time of 0.1 s. In the bilinear transformation with frequency prewarping, choose $\omega_c = 1$ rad/s, the half-power frequency of $G(s)$. Plot the frequency response of $G(s)$ and $G(z)$ for ω from 0.1 to 10 rad/s.

Solution

In Script 6.5, after the analog filter coefficients have been used to generate the continuous-time filter, the functions `imp_inv` and `c2d` are used to find the various discrete-time filters:

Impulse invariant:

$$G(z) = \frac{0.00931z}{z^2 - 1.859z + 0.8681}$$

Tustin:

$$G(z) = \frac{0.002329(z^2 + 2z + 1)}{z^2 - 1.859z + 0.8682}$$

Prewarping:

$$G(z) = \frac{0.002333(z^2 + 2z + 1)}{z^2 - 1.859z + 0.8681}$$

Matched-z:

$$G(z) = \frac{0.004659(z + 1)}{z^2 - 1.859z + 0.8681}$$

Note that the impulse-invariant and matched-z transformations have identical poles but different zeros. These poles are close but not identical to those of the Tustin and prewarped equivalents. The frequency response plots in parts (a) and (b) of Figure 6.5 show that all these filters provide a good approximation of the analog filter, at least up to 2 rad/s.

MATLAB Script

```
% Script 6.5:  Discrete equivalent transformations
Ts = 0.1                               % sampling time
Ga = tf(1,[1 sqrt(2) 1])               % analog (continuous-time) filter
w = logspace(-2.5,0,50)*pi/Ts;         % frequency points
[maga,pha] = bodedb(Ga,w);             % analog frequency response
Gii = imp_inv(Ga,Ts)                   % impulse invariant transf
[magi,phi] = bodedb(Gii,w);
Gtust = c2d(Ga,Ts,'tustin')            % Tustin (bilinear) rule
[magt,pht] = bodedb(Gtust,w);
Gprew = c2d(Ga,Ts,'prewarp',1)         % frequency prewarping
[magp,php] = bodedb(Gprew,w);
Gmat = c2d(Ga,Ts,'matched')            % matched z-transform
[magm,phm] = bodedb(Gmat,w);
%------ plot magitudes for all cases ------------
semilogx(w,maga,w,magi,'o',w,magt,'*',w,magp,'+',w,magm,'x'),grid
axis([w(1) w(end) -80 5])
%------ plot phases for all cases ------------
semilogx(w,pha,w,phi,'o',w,pht,'*',w,php,'+',w,phm,'x'),grid
axis([w(1) w(end) -270 0])
```

(a)

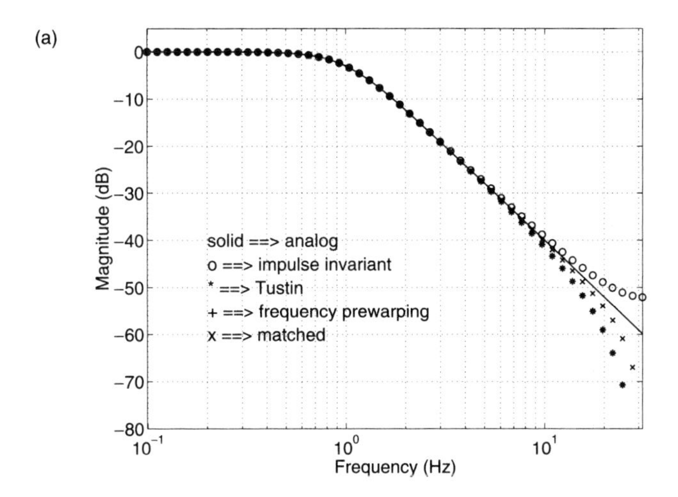

(b)

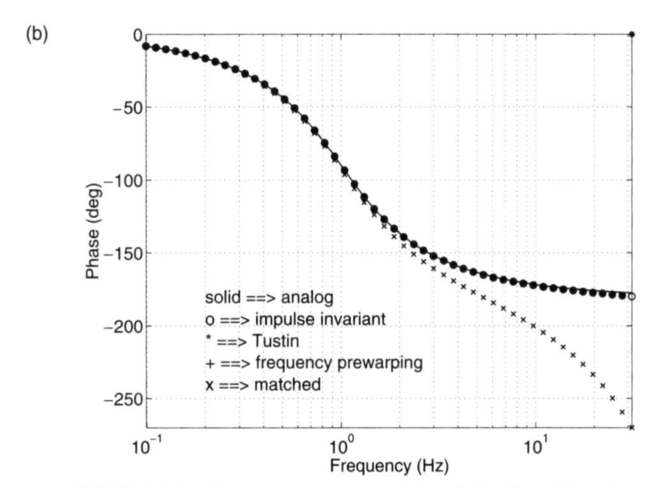

FIGURE 6.5 *Frequency response plots of the five filters for Example 6.5: (a) magnitude; (b) phase*

□

REINFORCEMENT PROBLEMS

P6.10 Second-order filter. Compute the various discrete equivalents for the analog filter with the transfer function

$$G(s) = \frac{s+3}{s^2+3s+2}$$

and compare the frequency responses of the filters using a sampling period of $T_s = 0.1$ s. For the prewarping design, use a center frequency of 1 rad/s.

P6.11 Bandstop Filter. Compute the various discrete equivalents for the analog filter whose transfer function is

$$G(s) = \frac{22500s}{s^4 + 212.132s^3 + 42500s^2 + 2121320s + 10^8}$$

and compare the frequency responses of the filters using a sampling period of $T_s = 0.001$ s. For the prewarping design, use a center frequency of 100 rad/s.

EXPLORATORY PROBLEMS

EP6.1 Geometric interpretation of frequency response. The frequency response of a discrete-time system $G(z)$ in the ZPK form

$$G(z) = K\frac{(z-z_1)\cdots(z-z_m)}{(z-p_1)\cdots(z-p_n)}$$

where $n \geq m$ and z_i and p_k are the zeros and poles, respectively, can be expressed as

$$G(\epsilon^{j\omega T_s}) = K\frac{M_1(\omega)\epsilon^{j\phi_1(\omega)}\ldots M_m(\omega)\epsilon^{j\phi_m(\omega)}}{N_1(\omega)\epsilon^{j\theta_1(\omega)}\ldots N_n(\omega)\epsilon^{j\theta_n(\omega)}}$$

where $M_i(\omega)$ and $\phi_i(\omega)$ are the magnitude and phase of the vector pointing from the zero z_i to the point $\epsilon^{j\omega T_s}$ on the unit circle, and $N_k(\omega)$ and $\theta_k(\omega)$ are the magnitude and phase of the vector pointing from the pole p_k to the point $\epsilon^{j\omega T_s}$. Thus the magnitude of the frequency response $G(\epsilon^{j\omega T_s})$ is given by

$$K(M_1(\omega)\cdots M_m(\omega))/(N_1(\omega)\cdots N_n(\omega))$$

and the phase is given by

$$(\phi_1(\omega) + \cdots + \phi_m(\omega)) - (\theta_1(\omega) + \cdots + \theta_n(\omega))$$

These geometric relationships give extra insights that help one visualize the frequency response of discrete-time systems.

These geometric relationships are utilized in the RPI M-file `dfilter.m` to generate the frequency response of a discrete-time system. This program allows the user to enter the poles and zeros of a digital filter via the keyboard or using the mouse to point to the locations. Then the frequency response is generated as a point tracing out the upper half of the unit circle, with vectors connecting to the poles and zeros. The gain of the digital filter is set such that the maximum gain is 0 dB. The user can interactively specify the digital filters, such as those in Examples 6.1 and 6.3, and examine their frequency response plots.

EP6.2 Finite-impulse response filters. Consider the nth-order digital filter whose transfer function is given by

$$G(z) = \frac{a_0 z^n + a_1 z^{n-1} + \cdots + a_n}{z^n} = a_0 + a_1 z^{-1} + \cdots + a_n z^{-n} \quad (6.7)$$

where all the poles of the filter are at the origin. Such a filter is called a *finite-impulse response (FIR) filter* because its impulse response, given by the sequence $[a_0\ a_1\ \ldots\ a_n]$, is of finite duration. [In contrast, filters such as (6.2) with poles not at the origin are called *infinite-impulse response (IIR) filters*.] FIR filters are commonly designed by applying windows to the impulse response of ideal filters. The interested reader is encouraged to consult Oppenheim and Schafer (1999) or similar textbooks on digital signal processing. Here we apply the procedure of polynomial long division to obtain an IIR filter from a FIR filter. Consider the transfer function of the bandpass filter $G(z)$ in Example 6.4 and perform a long division of the numerator by the denominator to obtain the infinite-series expansion

$$G(z) = a_0 + a_1 z^{-1} + \cdots + a_n z^{-n} + \cdots \quad (6.8)$$

Obtain three FIR filters, $G_2(z), G_4(z)$, and $G_6(z)$ by truncating $G(z)$ for $n = 2$, 8, and 14. Plot the magnitudes of the frequency responses of the resulting filters. Is $n = 14$ sufficient to closely approximate the bandpass characteristic of the original IIR filter, or do we need more terms?

Having completed the steps outlined above, apply the same technique to some other IIR filters that have been considered in this chapter.

COMPREHENSIVE PROBLEMS

CP6.1 Filtering of the drumbeat sound track. As discussed in Comprehensive Problem CP5.1, the drumbeat sound track has a noise component at about 7.1 kHz. Design a lowpass filter to reduce that component by 90%. Listen to the filtered sound track and check whether the filtering meets your expectation.

Many MATLAB sound tracks can be found in the `toolbox/matlab/datafun` subdirectory. Listen to a few of these sound tracks and investigate whether any filtering can improve the sound quality. You can also add some noise to the signals using the random number generator function `rand` and apply filters such as those from the MATLAB Signal Processing Toolbox. Listen to the input and output sound tracks. Then apply the `fft` function to determine and plot the frequency contents of these signals.

In addition, you can record a sound track from an audio tape that has a lot of hissing noise in the background. Apply lowpass filtering to investigate whether the sound quality can be improved. You can capture the sound track as a `.wav` file on your PC. MATLAB has a function, `wavread`, that will read these files into the MATLAB command window workspace.

CP6.2 2D deblurring. Load the `clown.mat` data file provided by MATLAB into the MATLAB command window workspace. Then the `image` function can be used to display a color picture of the clown, which is contained in an array `X` of the size 200 pixels by 320 pixels. A blurred image of the clown has been created and saved in the `clownb.mat` file by adding to the good image a weaker image of itself shifted 10 pixels left and up; that is, the blurred image is related to the good image by

$$X_b(i, j) = X(i, j) + \alpha X(i + 10, j + 10) \tag{6.9}$$

where $0 < \alpha < 1$. Dealing with only one axis of (6.9) and regarding a positive shift as a time-stage advance, we can apply the z-transform to (6.9) to obtain

$$X_b(z) = X(z) + \alpha z^{10} X(z) = (1 + \alpha z^{10}) X(z) \tag{6.10}$$

Given only $X_b(z)$, (6.10) shows that $X(z)$ can be recovered from

$$X(z) = \frac{1}{(1 + \alpha z^{10})} X_b(z) = (1 - \alpha z^{10} + \alpha^2 z^{20} - \cdots) X_b(z) \tag{6.11}$$

Apply (6.11) to both the x- and y-axes and try to recover the good image from the blurred image, using an FIR approximation of (6.11). You have to estimate α by trial and error. Can you return the clown's hair color to yellow and her nose color to red?

SUMMARY

In this chapter we have examined the frequency response of a discrete-time system and related it to digital filtering. Low-order digital filters were investigated, and the frequency response of a digital filter was used to derive the sinusoidal steady-state response of its output. Transformations to derive digital filters from analog filters were also presented and illustrated.

ANSWERS

P6.1 The poles of $G(z)$ are at $z = 0.4425\epsilon^{\pm j2.0015}$, and $G(z)$ has a double zero at $z = 1$.

P6.2 The poles of $G(z)$ are at $z = 0.8095\epsilon^{\pm j1.2406}$ and at $z = 0.8095\epsilon^{\pm j1.9010}$. $G(z)$ has a double zero at $z = 1$ and a double zero at -1.

P6.3 The poles of $G(z)$ are at $z = 0.6856\epsilon^{\pm j0.9919}$ and $z = 0.6856\epsilon^{\pm j2.1497}$. The zeros of $G(z)$ are at $z = 1.0\epsilon^{\pm j1.3449}$ and $z = 1.0\epsilon^{\pm j1.7967}$.

P6.4 $y_{1\text{ss}}(kT_s) = 0.0042\sin(4\pi kT_s + 196.29°)$,
$y_{2\text{ss}}(kT_s) = 1.4144\cos(60\pi kT_s - 55.0°)$

P6.5 $y_{1\text{ss}}(kT_s) = 0.0048\sin(4\pi kT_s + 196.6°)$,
$y_{2\text{ss}}(kT_s) = 1.8884\cos(60\pi kT_s - 215.10°)$

P6.6 $y_{1\text{ss}}(kT_s) = 2.00\sin(4\pi kT_s + 12.92°)$,
$y_{2\text{ss}}(kT_s) = 0.200\cos(60\pi kT_s - 1.87°)$

P6.7 The extreme values of the filter output are approximately ± 10. The magnitudes of the sinusoidal components in the filter's output are 0.6001 at 7 Hz, 1.515 at 25 Hz, and 4.021 at 43 Hz.

P6.8 The magnitude of the filter's frequency response is 21.35 dB at $\omega = 10.0$ rad/s, 0 dB at $\omega = 0.5\pi/T_s = 157.08$ rad/s, and 21.37 dB at $\omega = \pi/T_s = 314.16$ rad/s. During the interval $9.5 < t < 10$ s, the minimum output values are between -21 and -23, whereas the maximum output values are between 21 and 24.

P6.9 The extreme values of the filter output are approximately ± 100. The magnitudes of the sinusoidal components in the filter's output are 6.770 at 7 Hz, 1.514 at 25 Hz, and 45.315 at 43 Hz.

P6.10 Impulse-invariant transformation:

$$G(z) = \frac{0.1z(z-0.7326)}{z^2 - 1.724z + 0.7408}$$

Bilinear (Tustin) transformation:

$$G(z) = \frac{0.04978z^2 + 0.01299z - 0.0368}{z^2 - 1.723z + 0.7403}$$

Bilinear transformation with prewarping at 1 rad/s:

$$G(z) = \frac{0.04982z^2 + 0.01301z - 0.03682}{z^2 - 1.723z + 0.7401}$$

Matched-z transformation:

$$G(z) = \frac{0.09983z - 0.07396}{z^2 - 1.724z + 0.7408}$$

P6.11 Impulse-invariant transformation:

$$G(z) = \frac{1.046 \times 10^{-5}(z^3 - 0.06827z^2 - 0.9317z)}{z^4 - 3.77z^3 + 5.35z^2 - 3.389z + 0.8089}$$

Bilinear transformation:

$$G(z) = \frac{2.518 \times 10^{-6}(z^4 + 2z^3 - 2z - 1)}{z^4 - 3.771z^3 + 5.353z^2 - 3.392z + 0.8096}$$

Bilinear transformation with prewarping at 100 rad/s:

$$G(z) = \frac{2.524 \times 10^{-6}(z^4 + 2z^3 - 2z - 1)}{z^4 - 3.77z^3 + 5.352z^2 - 3.391z + 0.8095}$$

Matched-z transformation: Because the filter has zero DC gain, the gain of $G(z)$ is adjusted to match the filter's maximum gain:

$$G(z) = \frac{5.051 \times 10^{-6}(z^3 + z^2 - z - 1)}{z^4 - 3.77z^3 + 5.35z^2 - 3.389z + 0.8089}$$

7 *System Performance*

PREVIEW

The main objectives of feedback control are to achieve closed-loop system stability and to satisfy certain desired performance criteria. As in a continuous-time system, the performance of a discrete-time system can be measured in either the time domain or the frequency domain. Common time-domain measures include the damping ratios of the dominant poles and the step-response rise time, overshoot, settling time, and steady-state command tracking error. For discrete-time systems, the damping ratios of the dominant poles can be computed by transforming them back to equivalent continuous-time system poles. In the frequency domain, typical performance measures include the gain and phase margins, DC gain, and closed-loop system bandwidth. The purpose of this chapter is to use MATLAB to evaluate system performance measures and to understand how the controller parameters may affect system performance.

TIME-DOMAIN PERFORMANCE

Controllers for feedback systems are designed to improve system performance so that certain desired criteria are satisfied. Common measures of closed-loop system performance include damping ratios and step-response quantities such as the rise time and overshoot. This section demonstrates the use of some of the MATLAB and RPI functions in computing system performance measures.

DAMPING RATIO

For a continuous-time system with the complex poles $s = \sigma \pm j\omega$, the damping ratio is defined as $\zeta = \cos\phi$, where $\phi = \tan^{-1}(-\omega/\sigma)$. To compute the damping ratio of the poles $z = \sigma_1 \pm j\omega_1 = r\epsilon^{\pm j\theta}$ of a discrete-time system with a sampling time T_s, we first use the impulse-invariant transformation $z = \epsilon^{sT_s}$ to transform them to the equivalent s-plane poles. Setting $r\epsilon^{\pm j\theta} = \epsilon^{(\sigma \pm j\omega)T_s} = \epsilon^{\sigma T_s}\epsilon^{\pm j\omega T_s}$, we obtain $\sigma = \ln(r)/T_s$ and $\omega = \theta/T_s$. Then the damping ratio of $z = \sigma_1 \pm j\omega_1 = r\epsilon^{\pm j\theta}$ can be computed from the equivalent s-plane poles $s = \sigma \pm j\omega$. This sequence of computation is performed in MATLAB by applying the `damp` function to the LTI system object, as shown in the following example, which is a follow-up to Example 5.8.

□ EXAMPLE 7.1 *Damping Ratio*

Compute the damping ratios of the poles of the closed-loop system considered in Example 5.8, where

$$G_p(s) = \frac{1}{s(s+1)}, \quad K(z) = 5\left(\frac{z-0.9}{z-0.7}\right), \quad \text{and} \quad H(s) = 1$$

and $T_s = 0.1$ s. Use the MATLAB `damp` function and verify the result using the transformation $z = \epsilon^{sT_s}$.

Solution

In Example 5.8, the discrete-time closed-loop system was shown to have the transfer function

$$T_a(z) = \frac{0.02419z^2 + 0.001626z - 0.02105}{z^3 - 2.581z^2 + 2.240z - 0.6544}$$

which has a real pole at $z = 0.8960$ and a pair of complex poles at $z = 0.8423 \pm j0.1445 = 0.8546\epsilon^{\pm j0.1699}$. The damping ratios of these three poles are computed by the `damp` function, as shown in Script 7.1, and the results are summarized in Table 7.1. For a positive real pole inside the unit circle, the damping ratio is always unity. For the complex z-plane poles, the equivalent s-plane poles are computed to be $s = -1.571 \pm j1.699$, indicating a damping ratio of 0.679 and an undamped natural frequency of 2.314 rad/s, which is simply the magnitude of the s-plane poles.

z-plane pole	Magnitude	Equivalent damping	Equivalent frequency (rad/s)
0.8960	0.8960	1.00	1.10
$0.8546\epsilon^{j0.1699}$	0.8546	0.679	2.314
$0.8546\epsilon^{-j0.1699}$	0.8546	0.679	2.314

TABLE 7.1 *Damping ratios of closed-loop system poles in Example 7.1*

MATLAB Script

```
% Script 7.1:  Damping ratio computation
Ts = 0.1                              % sampling time
Gp = tf(1,[1 1 0])                    % plant in continuous TF form
Gz = c2d(Gp,Ts,'zoh')                 % discretization with zero-order hold
Kz = tf(5*[1 -0.9],[1 -0.7],Ts)       % controller in discrete TF form
Ta = feedback(Gz*Kz,1,-1)             % CL transfer function
p = pole(Ta)                          % z-plane poles of closed-loop system
[r,del] = xy2p(p)                     % z-plane poles in polar form
damp(Ta)                              % damping ratios
sig = log(r)/Ts                       % real part of equiv. s-plane pole
w = del/Ts                            % imag part of equiv. s-plane pole
zeta = cos(atan(-w./sig))             % damping ratio
wn = sqrt(sig.^2+w.^2)                % undamped natural frequency
```

□

S-PLANE TO Z-PLANE MAPPING

To more explicitly illustrate the pole transformation $z = \epsilon^{sT_s}$ and to gain further insight into the correspondence between the s- and z-plane poles, we will select loci of poles in the s-plane having specific features and transform them into the corresponding loci in the z-plane. These loci include:

- constant real part
- constant imaginary part
- constant damping ratio
- constant undamped natural frequency

The following example illustrates the mapping of these loci. In this example, we also illustrate the use of the `sgrid` and `zgrid` commands, which plot lines of constant damping ratios and undamped natural frequencies on the s- and z-plane, respectively.

☐ EXAMPLE 7.2 *Mapping S-Plane to Z-Plane Poles*

Assume a sampling time of $T_s = 0.1$ s and map the following sets of s-plane loci to the z-plane, using the transformation $z = \epsilon^{sT_s}$.

a. the horizontal lines $\sigma + j\omega$ for $-35 \leq \sigma \leq 0$, and $\omega = 0$, $0.25\pi/T_s$, $0.5\pi/T_s$, $0.75\pi/T_s$, and π/T_s rad/s

b. the vertical lines $\sigma + j\omega$ for $\sigma = 0$, -5, -10, -20, and -30, and $0 \leq \omega \leq \pi/T_s$ rad/s

c. the constant damping ratio lines for angles of $\phi = 0$, 22.5, 45, 67.5, and 90 degrees clockwise from the negative real axis

d. the constant undamped natural frequency lines for $\omega_n = 0$, $0.25\pi/T_s$, $0.5\pi/T_s$, $0.75\pi/T_s$, and π/T_s rad/s

Solution

In Script 7.2(a), we generate the horizontal line on the negative real axis of the s-plane by defining the variable `s0` for a set of discrete values of σ going from 0 to -35 in 20 equal decrements. To generate the other horizontal lines, we merely add the appropriate imaginary part to `s0`. For example, the middle horizontal line is given by `s0 + j*0.5*pi/Ts`. For expediency, we generate these five lines in a single MATLAB command and store them as a single array, `s`. These s-plane lines are mapped into the z-plane according to the expression $z = \epsilon^{sT_s}$. The command `z = exp(s*Ts)` performs this task by generating an array, `z`, with the same dimension as `s`, whose (p, q) entry is the exponent of the (p, q) entry of `s` multiplied by T_s.

The s-plane loci for part (a) of Example 7.2 are shown in Figure 7.1(a), and the corresponding z-plane loci, which are uniformly-spaced rays emanating from the unit circle (for $\sigma = 0$) and approaching the origin as σ becomes more negative, are shown in Figure 7.1(b). The corresponding curves are marked by the same symbols. In particular, the s-plane negative real axis maps onto the z-plane positive real axis, and the line whose imaginary part is half of the sampling frequency in the s-plane maps onto the negative real axis of the z-plane. By matching up the plotting symbols for the lines, we see that the points on the negative real axis in the s-plane map into points on the positive real axis in the z-plane, on or inside the unit circle.

Note that we show the mapping only into the upper half of the unit circle in the z plane. Mapping into the lower half of the unit circle can be obtained from the complex conjugates of these s-plane curves.

MATLAB provides the `sgrid` and `zgrid` functions to mark the constant damping ratio and natural frequency lines in the s- and z-planes, respectively. In the s-plane, the constant damping ratio lines are in increments of 0.1, with the imaginary axis having a damping ratio of 0, and the negative real axis having a damping ratio of 1. In the z-plane, the constant damping ratio lines are also in increments of 0.1, with the unit circle having a damping ratio of 0, and the constant natural frequency lines are in increments of 1/10 of half of the sampling frequency.[1] These functions will be applied to appropriate plots in this example to highlight how damping ratios and natural frequencies vary as the branches vary.

[1] To plot a single constant damping ratio line, the RPI function `dzline` can be used.

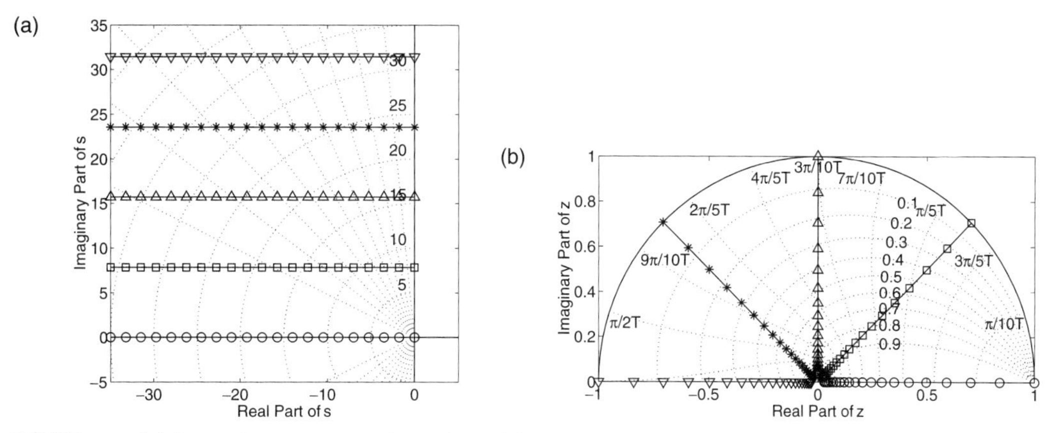

FIGURE 7.1 *(a) Lines of constant ω in the s-plane; (b) Mapping of lines of constant ω in the z-plane*

MATLAB Script

```
% Script 7.2(a):  Mapping of horizontal s-plane lines into the z-plane
Ts = 0.1                              % sampling time
% ------------ set up s-plane points -------------
xx = [0:0.05:1]'                      % set of values from 0...1
N = length(xx)                        % number of points in xx
s0 = -xx*35                           % negative real axis
% generate an array, each column is a horizontal line;
% the real part of each column is identical
s = s0*[1 1 1 1 1] + j*ones(N,1)*[0 0.25 0.5 0.75 1]*pi/Ts;
% draw s-plane loci - also for Scripts 7.2(b), (c), and (d)
plot(real(s(:,1)),imag(s(:,1)),'-o',real(s(:,2)),imag(s(:,2)),'-s',...
     real(s(:,3)),imag(s(:,3)),'-^',real(s(:,4)),imag(s(:,4)),'-*',...
     real(s(:,5)),imag(s(:,5)),'-v'), sgrid
% conversion to z-plane poles - also for Scripts 7.2(b), (c), and (d)
z = exp(s*Ts);
% draw z-plane loci
plot(real(z(:,1)),imag(z(:,1)),'-o',real(z(:,2)),imag(z(:,2)),'-s',...
     real(z(:,3)),imag(z(:,3)),'-^',real(z(:,4)),imag(z(:,4)),'-*',...
     real(z(:,5)),imag(z(:,5)),'-v'), zgrid
```

For part (b) of Example 7.2, we can create a set of points on the positive imaginary axis in the s-plane and then shift them to the left by adding the same negative real part to each point. The plotting and transformation commands are essentially the same as those for part (a). So in Script 7.2(b), we show only those commands required to generate the vertical lines in the s-plane, with all the vertical lines contained in a single array, s. The plots produced by the completed script are shown in Figure 7.2. Observe that the positive imaginary axis in the s-plane maps onto the unit circle. In addition, a vertical line in the left half of the s-plane maps into a circle centered at the origin and inside the unit circle. The z-plane circles become smaller as the s-plane vertical lines move farther away from the imaginary axis.

MATLAB Script

```
% Script 7.2(b):  Mapping of vertical s-plane lines into the z-plane
Ts = 0.1                          % sampling time
xx = [0:0.05:1]'                  % set of values from 0...1
N = length(xx)                    % number of points in xx
s0 = j*xx*pi/Ts                   % imaginary axis
% generate an array, each column is a vertical line
s = ones(N,1)*[0 -5 -10 -20 -30] + s0*[1 1 1 1 1];
```

For part (c) of Example 7.2, we need to create lines of constant damping ratio ζ in the s-plane. Although this can be done in a variety of ways, we will start with the points in `s0` in Script 7.2(b) that lie on the positive imaginary axis of the s-plane. Then we will add to these points a negative real amount proportional to the imaginary part of the point. In Script 7.2(c), we show only those commands required to generate the straight lines of constant damping ratio in the s-plane. The lines contained in the array `s` have damping ratios of 0, 0.383, 0.707, 0.924, and 1. The plots produced by the completed script are shown in Figure 7.3. Note that the s-plane imaginary axis points (circles), which correspond to zero damping, map into the unit circle of the z-plane, and the lines corresponding to positive damping map into logarithmic spirals that start at $z = 1$ and approach $z = 0$.

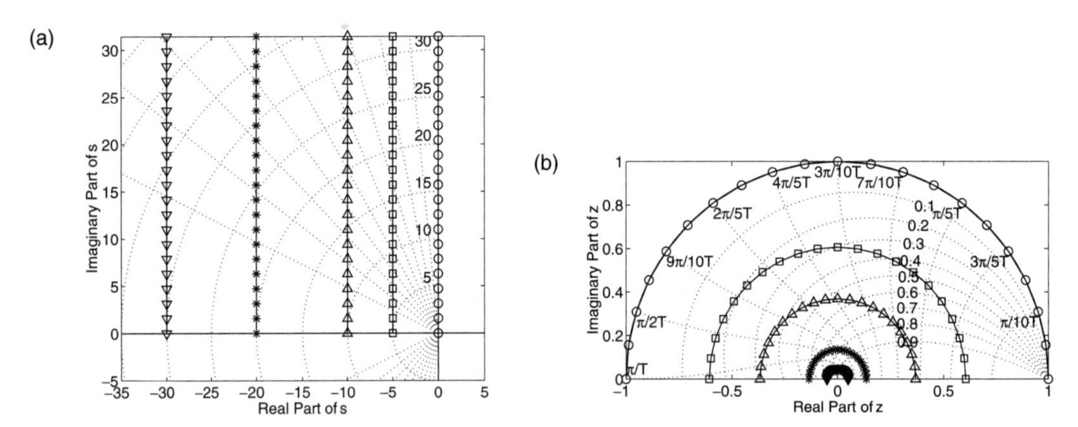

FIGURE 7.2 *(a) Lines of constant σ in the s-plane; (b) Mapping of lines of constant σ in the z-plane*

MATLAB Script

```
% Script 7.2(c):  Mapping of constant damping ratio s-plane lines into z-plane
Ts = 0.1                                  % sampling time
xx = [0:0.05:1]'                          % set of values from 0...1
N = length(xx)                            % number of points in xx
s0 = j*xx*pi/Ts                           % positive imaginary axis
% generate an array, each column is a constant damping ratio line
% cos(67.5 deg) = 0.383, cos(45 deg) = 0.707, cos(22.5 deg) = 0.924
s = s0*[1 1 1 1] - imag(s0)*[0  1/tan(67.5*pi/180) ...
    1/(tan(45*pi/180))  1/(tan(22.5*pi/180))];
s = [s real(s(:,3))];                     % add negative real axis to the array
```

For part (d) of Example 7.2, the s-plane loci are arcs of constant undamped natural frequency ω_n. First we define a set of points that lie on the arc of radius π/T_s in the upper-left quadrant of the s-plane. This can be done by defining a set of equally spaced angles, ϕ_i, between 0 and 90 degrees. Then the quantities $-\cos\phi_i + j\sin\phi_i$ will have unit length and lie in the upper-left quadrant. To get an arc of specific radius, we just multiply these complex quantities by the desired radius. The code fragment in Script 7.2(d) accomplishes this. When the appropriate transformation and plotting commands are added [see Script 7.2(a)], the plots in parts (a) and (b) of Figure 7.4 result. Note that the s-plane arc of 0 radius is the point $s = 0$, which maps into the point $z = 1$.

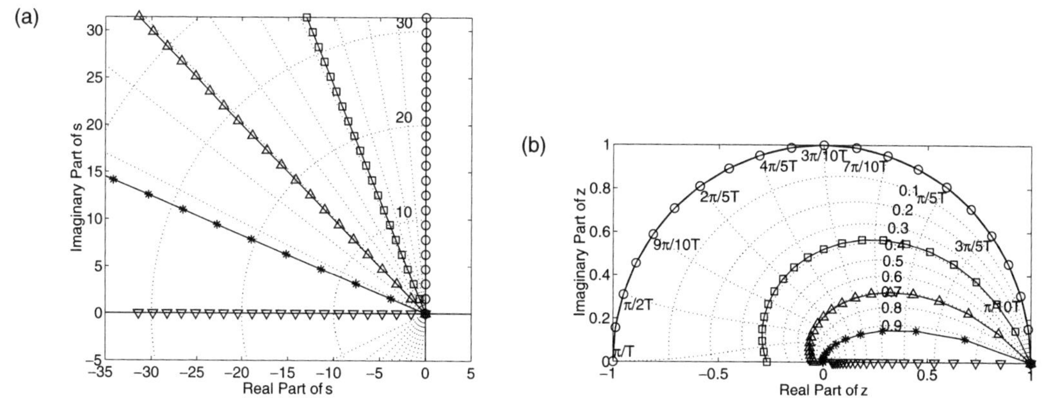

FIGURE 7.3 *(a) Lines of constant damping ratio ζ in the s-plane; (b) Mapping of lines of constant damping ratio ζ in the z-plane*

MATLAB Script

```
% Script 7.2(d):  Mapping of constant Wn s-plane lines into the z-plane
Ts = 0.1                                  % sampling time
xx = [0:0.05:1]'                          % set of values from 0...1
N = length(xx)                            % number of points in xx
phi = xx*pi/2                             % equal angles from 0 --> 90 deg
% generate an array, each column is a constant frequency line
s0 = (pi/Ts)*(-cos(phi)+j*sin(phi));      % arc of radius pi/Ts
s = s0*[1 0.75 0.5 0.25 0];
```

(a)

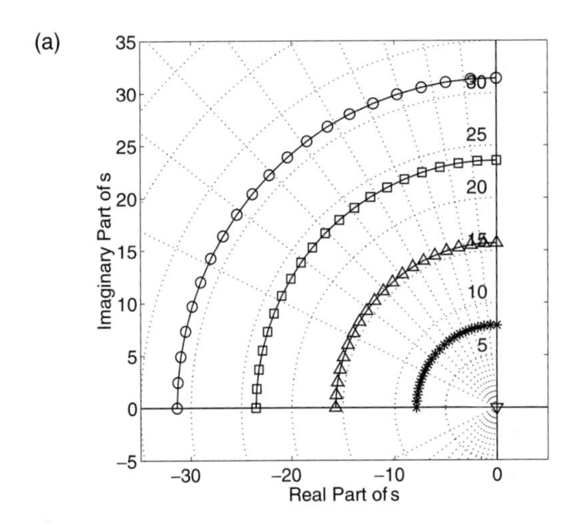

(b)

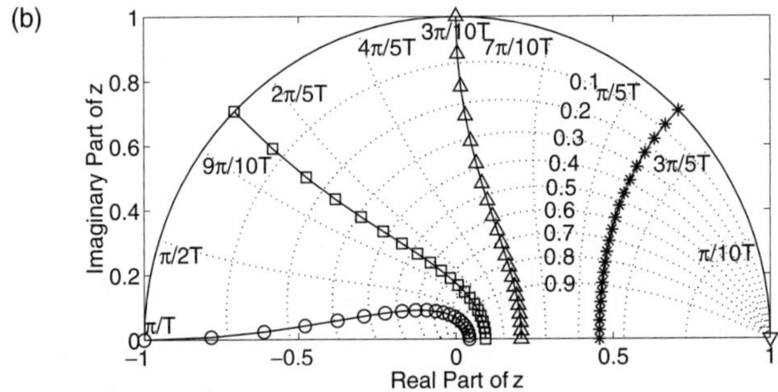

FIGURE 7.4 *(a) Lines of constant natural frequency ω_n in the s-plane; (b) Mapping of lines of constant natural frequency ω_n in the z-plane*

WHAT IF? The s-plane pole loci of Example 7.2 are all in the left half-plane (LHP), resulting in the z-plane pole loci being inside the unit circle. Now extend these s-plane pole loci into the right half-plane (RHP) and plot the corresponding z-plane pole loci. They should all be outside of the unit circle in the z-plane. What else do you observe? ■ □

STEP RESPONSE

In the time domain, the command-tracking performance of the controller is commonly measured in terms of how well the closed-loop system response $y(kT_s)$ follows a reference step input $r(kT_s) = y_{\text{ref}}$. The performance measures include t_r, the rise time for $y(kT_s)$ to go from 10% to 90% of the reference value y_{ref}; M_o, the maximum overshoot in percent, defined as

$$M_o = \frac{y_{\text{peak}} - y_{\text{ref}}}{y_{\text{ref}}} \times 100$$

where y_{peak} is the peak value of $y(kT_s)$; t_s, the settling time for $y(kT_s)$ to remain within 2% of y_{ref}; and e_{ss}, the steady-state regulation error.

The RPI function `kstats` has been created to compute these performance measures. The arguments of `kstats` are (i) a column vector of time values, (ii) a corresponding column vector of the step response to be evaluated, and (iii) a scalar specifying the reference value of y_{ref}. The use of `kstats` is illustrated in the following example.

□ EXAMPLE 7.3 *Step-Response Performance*

For the closed-loop systems in Example 5.8, use `kstats` to find the quantities M_o, t_p, t_r, t_s, and e_{ss}. Compare the performance measures of the system with the ideal sensor with those of the system with a slow sensor, referring to the step-response plots computed in Example 5.8 and shown in Figure 5.11.

Solution

The closed-loop system models $T_a(z)$ and $T_b(z)$ from Example 5.8 are computed and used to generate the step responses. The step responses, together with the time array and the reference value (set to unity), are used as the inputs to the `kstats` function. The computed performance measures are listed in Table 7.2.

The performance measures show clearly the effect of the slow sensor in the closed-loop system. With the slow sensor, it takes longer for the feedback signal to affect the control signal, resulting in a faster initial response (shorter time constant). This leads to a higher overshoot, however, which is less desirable. Note that the simulation has not been run long enough for the values of the steady-state error shown in the table to equal their theoretical values of zero, although they are very close. The step responses shown in Figure 5.11 and reproduced in Figure 7.5, are consistent with these conclusions.

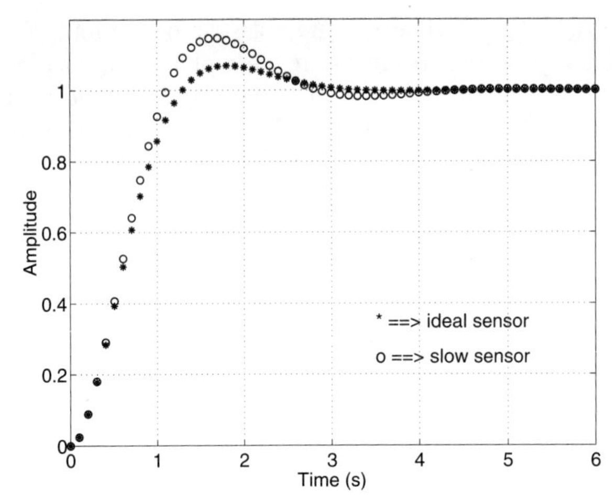

FIGURE 7.5 *Step responses with the ideal sensor and the slow sensor for Example 7.3*

Sensor	*Rise time* t_r (s)	*Overshoot* M_o (%)	*Peak time* t_p (s)	*Settling time* t_s (s)	*Steady-state error* e_{ss} (%)
Ideal	0.859	6.87	1.80	2.70	0.02
Slow	0.756	14.69	1.70	2.70	0.04

TABLE 7.2 *Time-domain performance measures for Example 7.3*

MATLAB Script

```
% Script 7.3:  Step response measures
Ts = 0.1                                % sampling time
Gp = tf(1,[1 1 0])                      % plant in continuous tf form
Gz = c2d(Gp,Ts,'zoh')                   % discretize using zero-order hold
Gc = tf(5*[1 -0.9],[1 -0.7],Ts)         % discrete controller in tf form
%-------- Ideal sensor ----------------
Ta = feedback(Gz*Gc,1,-1)               % CL transfer function
k  = [0:1:60]';                         % discrete-time index
ya = step(Ta,k*Ts);                     % step response
[Moa,tpa,tra,tsa,essa] = kstats(k*Ts,ya,1)
%-------- Slow sensor -----------------------
H  = tf(10,[1 10])                      % continuous sensor in tf form
HGz = c2d(H*Gp,Ts,'zoh')                % discretize combined H & Gp
Tb = Gz*Gc/(1+HGz*Gc)                   % CL transfer fucntion
Tb = minreal(Tb)                        % eliminate duplicate states
yb = step(Tb,k*Ts);                     % step response
[Mob,tpb,trb,tsb,essb] = kstats(k*Ts,yb,1)
```

□

REINFORCEMENT PROBLEMS

For the following closed-loop systems, find the damping ratios and natural frequencies of the closed-loop system poles using `damp`, and the performance measures M_o, t_p, t_r, t_s, and e_{ss} of the closed-loop system step response using `kstats`.

P7.1 ZPK model with $H(s) = 1$. Use the closed-loop system from Problem 5.12.

P7.2 State-space model with $H(s) = 1$. Use the closed-loop system from Problem 5.13.

P7.3 ZPK model with non-ideal sensor. Use the closed-loop system from Problem 5.14. Compare the performance of the closed-loop system with the performance measures obtained in Problem 7.1.

P7.4 State-space model with non-ideal sensor. Use the closed-loop system from Problem 5.15. Compare the performance of the closed-loop system with the performance measures obtained in Problem 7.2.

STEP RESPONSE MEASURES AS FUNCTIONS OF CONTROLLER PARAMETERS

The performance of a feedback system depends on the type and parameters of its controller. For example, when a first-order controller is used, the feedback system performance depends in a nonlinear manner on three parameters: the gain, the zero, and the pole. Using MATLAB, these nonlinear relationships can be plotted and used to select parameter values that satisfy design specifications. To illustrate this technique, we revisit Example 7.1, varying two controller parameters and using MATLAB's 3D plotting functions `mesh` and `contour` for visualization of the nonlinear dependency.

☐ EXAMPLE 7.4 *Parametric Variations*

For the system $G_p(s) = 1/[s(s+1)]$ in Example 7.1, we parameterize the controller as

$$G_c(z) = K\left(\frac{z - z_0}{z - 7z_0/9}\right)$$

so that the controller has only two design parameters, K and z_0. Vary the gain K from 2 to 10 in increments of 0.5 and z_0 from 0.1 to 0.9 in increments of 0.05. For each set of K and z_0, perform a step response and find the rise time t_r and the percent overshoot M_o. Plot these two values versus K and z_0 and find the values to achieve a rise time of not more than 0.6 s and an overshoot of not more than 20%. Use $H(s) = 1$ and $T_s = 0.1$ s.

Solution

The discretization of the sampled-data system is performed in part (a) of Script 7.4. In part (b), the arrays `z0` and `K` are set up to store the range of parameter variations. The variables `tr_mat` and `Mo_mat` are temporarily set to be zero matrices of appropriate dimensions. For each set of values for `z0` and `K`, the

MATLAB Script

```
% Script 7.4:  Performance measures as functions of control parameters
%--------- Part a:  discretization ----------------
Ts = 0.1                                 % sampling time
Gp = tf(1,[1 1 0])                       % plant in continuous tf form
Gz = c2d(Gp,Ts,'zoh')                    % discretize using zero-order hold
%--------- Part b:  set up parameter values ------------------
z0 = 0.1:0.05:0.9;                       % range of zeros
K  = 2:0.5:10;                           % range of gains
nz0 = length(z0)                         % number of values in z0
nK = length(K)                           % number of values in K
tr_mat = zeros(nz0,nK);                  % array to store rise time values
Mo_mat = zeros(nz0,nK);                  % array to store overshoot values
dtime  = [0:1:60]'*Ts;                   % time array from 0 to 6 s
%--- Part c:  for loops to apply various control parameter values -----
for ii = 1:nz0
 for jj = 1:nK
  Gcz = tf(K(jj)*[1 -z0(ii)],[1 -7/9*z0(ii)],Ts); % controller
  T = feedback(Gz*Gcz,1,-1);             % CL transfer function
  y = step(T,dtime);                     % step response
  [Mo,tp,tr,ts,ess] = kstats(dtime,y,1);  % find overshoot and rise time
  tr_mat(ii,jj) = tr;                    % store rise time for 3D plotting
  Mo_mat(ii,jj) = Mo;                    % store overshoot for 3D plotting
 end
end
%----------- Part d:  3D plotting -----------------------
mesh(K,z0,tr_mat)                        % 3D plot of rise time
xlabel('Gain')                           % label for x axis
ylabel('Zero location')                  % label for y axis
zlabel('Rise time (s)')                  % label for z axis
mesh(K,z0,Mo_mat)                        % 3D plot of overshoot
xlabel('Gain')                           % label for x axis
ylabel('Zero location')                  % label for y axis
zlabel('Percent overshoot')              % label for z axis
tr_vals = [0.3 0.4 0.5 0.6 0.7 0.8 1.0 1.2]  % rise-time contour values
[ctr,htr] = contour(K,z0,tr_mat,tr_vals);% rise-time contour lines
clabel(ctr,htr)                          % label rise-time contour lines
ylabel('Zero location'), hold on
Mo_vals = [10 20 30 40 50 60 70]         % percent overshoot contour values
[cMo,hMo] = contour(K,z0,Mo_mat,Mo_vals,'--');% overshoot contour lines
clabel(cMo,hMo)                          % label overshoot contour lines
plot(7.0, 0.87,'x')                      % X marks a design solution
set(findobj('marker','x'),'markersize',12)    % ...satisfying Mo<=20 pct
set(findobj('marker','x'),'linewidth',2.0)    % ...& tr <=0.6 s
hold off
```

closed-loop system transfer function is computed. As shown in part (c), this requires setting up a nested **for** loop. The step response for each closed-loop system is simulated to obtain the rise time `tr` and the percent overshoot `Mo` using the **kstats** function, and stored in the appropriate locations in `tr_mat` and `Mo_mat`.

The MATLAB command `mesh(K,z0,tr_mat)` plots the rise-time values in `tr_mat` in the z-axis given the x-axis coordinates in `K` and the y-axis coordinates in `z0`, as shown in Figure 7.6(a). Note that the columns of `tr_mat` are the x-axis variables `K` and the rows are the y-axis variables `z0`. Similarly, we can do a **mesh** plot for the percent overshoot, as shown in Figure 7.6(b). The mesh plots offer an overall visual perspective but cannot readily be used to read off parameter values.

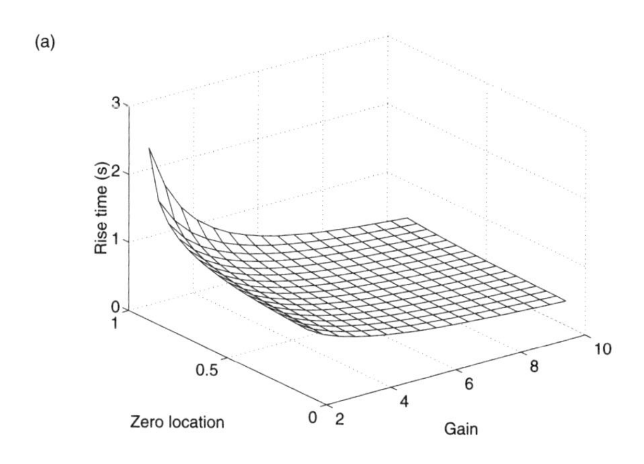

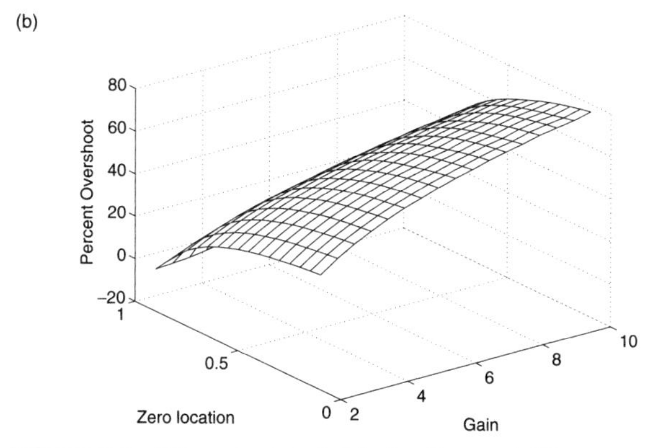

FIGURE 7.6 *3D mesh plot of (a) rise time and (b) percent overshoot, as functions of controller gain K and zero location z_0*

With two parameters, a contour plot of equal performance values is more useful. In MATLAB, the command `contour(K,z0,tr_mat,tr_vals)`, whose first three input arguments are identical to those of the corresponding `mesh` command, produces the solid curves in Figure 7.7. The fourth argument, `tr_vals`, defines the rise-time values for which the contour lines are drawn. The labels on the contour can be obtained by applying the command `clabel(ctr,htr)`, using the variables `ctr` containing the contour line information and `htr` containing the handle graphics data for each of the contour lines, which are the outputs of the `contour` function. The dashed contour lines for the percentage overshoot values are superimposed on the rise-time contour lines in Figure 7.7. These contour lines readily provide the information needed to select controller parameters to satisfy the required performance specifications. For example, to achieve a rise time of not more than 0.6 s and an overshoot of not more than 20%, we can use $K = 7.0$ and $z_0 = 0.87$, marked by the "×" in Figure 7.7.

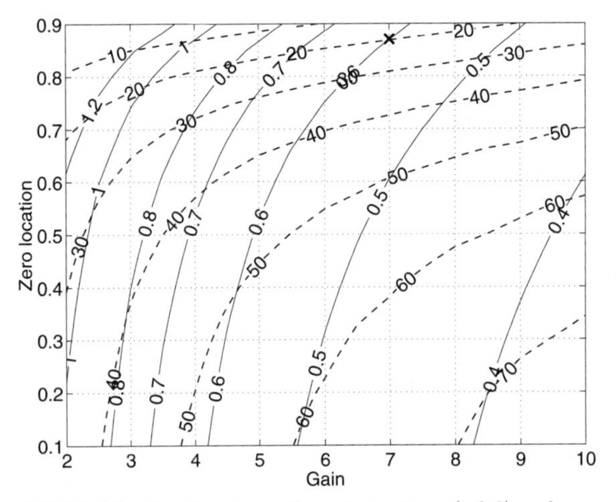

FIGURE 7.7 *Level contour plots of rise time (solid) and percent overshoot (dashed) of the closed-loop system, as functions of controller gain K and zero location z_0*

□

STEADY-STATE REGULATION

Consider the sampled-data feedback control system given in Figure 5.10 with $H(s) = 1$. The discrete equivalent of the system is given in Figure 7.8, where $G(z) = \mathcal{Z}\{G_p(s)G_{h0}(s)\}$.

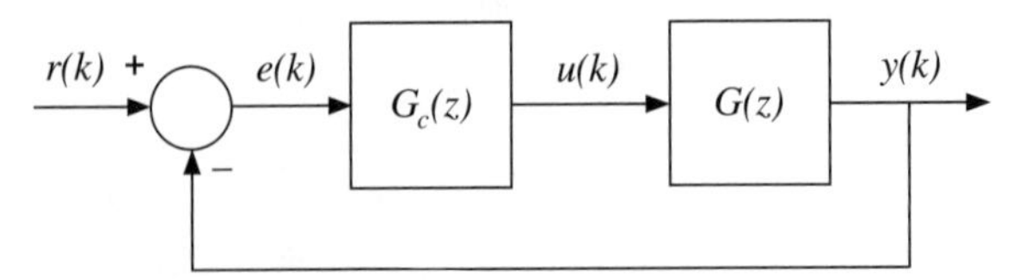

FIGURE 7.8 *Discrete equivalent of sampled-data feedback control system*

To analyze the steady-state tracking of the reference input $r(k) = y_{\text{ref}}$ by the plant output $y(k)$, it is useful to introduce the concept of system type. The open-loop system $G(z)G_c(z)$ is classified as a type-N system if $G(z)G_c(z)$ has N poles at $z = 1$. For a type-0 system, we define the *position error constant* as

$$K_p = \lim_{z \to 1} G(z)G_c(z) \tag{7.1}$$

If the closed-loop system is asymptotically stable, the steady-state error for a step input, $r(k) = AT_s$, is $e_{\text{ss}} = AT_s/(1 + K_p)$. For a type-1 system, we define the *velocity error constant* as

$$K_v = \lim_{z \to 1} (z - 1)G(z)G_c(z) \tag{7.2}$$

If the closed-loop system is asymptotically stable, the steady-state error for a ramp input, $r(k) = BkT_s$, is $e_{\text{ss}} = BT_s/K_v$.

To compute K_p, the Control System Toolbox provides the `dcgain` function, the usage of which has already been illustrated in several examples. It does not have a function to compute K_v directly, however. Therefore, we have created the RPI function `vgain` for this purpose. Given the discrete-time LTI object `G`, the `Kv = vgain(G)` command will return the velocity error constant K_v, as illustrated in the following example.

☐ EXAMPLE 7.5 *Position and Velocity Error Constants*

Determine the type of the system in Example 7.3 with $H(s) = 1$ and find K_p and K_v. Simulate the unit-ramp response of the closed-loop system.

Solution

The open-loop system in Example 7.3 is of type 1. This can be seen by observing either that the continuous-time plant $G_p(s)$ has a single pole at $s = 0$ (an integrator), or that the discretized plant

$$\begin{aligned} G(z) = \mathcal{Z}\{G_p(s)G_{h0}(s)\} &= \frac{0.004837z + 0.004679}{z^2 - 1.905z + 0.9048} \\ &= \frac{0.004837(z + 0.9672)}{(z - 1)(z - 0.9048)} \end{aligned}$$

obtained from Script 7.5 has a single pole at $z = 1$. To find K_p, we apply the `dcgain` function to the series connection $G(z)G_c(z)$, which, as expected, returns a value of infinity. To find K_v, we apply the `vgain` function to the same series connection, obtaining the velocity error constant $K_v = 0.1667$. Thus the steady-state error due to a unit-ramp input is $0.1/0.1667 = 0.60$, as can be seen in the unit-ramp response of the closed-loop system shown in Figure 7.9.

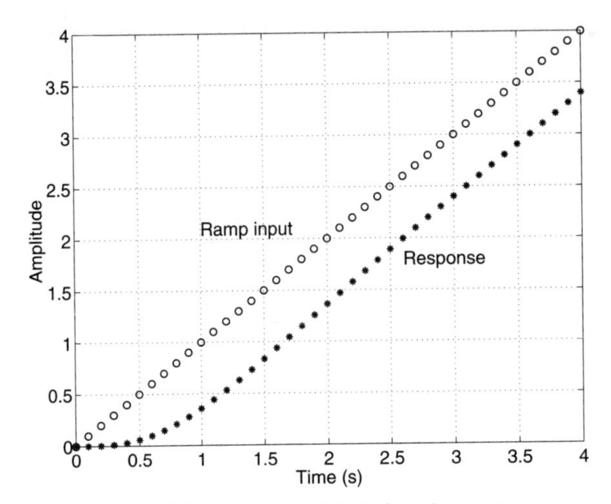

FIGURE 7.9 *Unit-ramp input (circles) and output response (asterisks) for Example 7.5*

MATLAB Script

```
% Script 7.5:  velocity error constant calculation
Gp = tf(1,[1 1 0])                     % plant in continuous tf form
Ts = 0.1                               % sampling time
Gz = c2d(Gp,Ts,'zoh')                  % discretization using zero-order hold
Gc = tf(5*[1 -0.9],[1 -0.7],Ts)        % controller in discrete tf form
Kp = dcgain(Gz*Gc)                     % position error constant
Kv = vgain(Gz*Gc)                      % velocity error constant
Ta = feedback(Gz*Gc,1,-1)              % CL transfer function
k  = [0:1:40]';
y  = lsim(Ta,k*Ts,k*Ts);               % simulate system with ramp input
plot(k*Ts,k*Ts,'o'),grid, hold on      % plot ramp input
plot(k*Ts,y,'*'), hold off             % plot ramp response
```

□

REINFORCEMENT PROBLEMS

In each of the following problems, consider the command tracking by the plant output for the sampled-data control system in Figure 5.10. The continuous-time plant $G_p(s)$ and the discrete-time controller $G(z)$ transfer function are given, and $H(s) = 1$ is assumed. Verify the system type number and find the appropriate steady-state error constant. Simulate the unit-step and unit-ramp responses of the closed-loop system.

P7.5 Type-0 system.

$$G_p(s) = \frac{6(s+10)}{(s+3)(s^2+4s+4)}, \quad G_c(z) = 0.5, \quad \text{and} \quad T_s = 0.1 \text{ s}$$

P7.6 Type-1 system.

$$G_p(s) = \frac{6(s+1)}{s(s^2+4s+4)}, \quad G_c(z) = 2.5, \quad \text{and} \quad T_s = 0.1 \text{ s}$$

FREQUENCY-DOMAIN PERFORMANCE

Open- and closed-loop system frequency responses provide additional means of evaluating the performance of sampled-data feedback systems. The gain and phase margins computed from the frequency response of the open-loop transfer function $G(z)G_c(z)$ provide an indication of the robustness of the sampled-data feedback system. Making use of the identity $\epsilon^{-j\pi} = -1$, we rewrite the characteristic equation $1 + G(z)G_c(z) = 0$ of the closed-loop system in Figure 7.8 as

$$G(z)G_c(z) = -1 = \epsilon^{-j\pi}$$

A stability requirement for the closed-loop system is that for all points on the unit circle ($z = \epsilon^{j\omega T_s}$), the function $G(z)G_c(z)$ must not be equal to -1. In particular, at the gain-crossover frequency ω_g, where $|G(\epsilon^{j\omega_g T_s}) G_c(\epsilon^{j\omega_g T_s})| = 1$, the phase of $G(\epsilon^{j\omega_g T_s})G_c(\epsilon^{j\omega_g T_s})$ should be greater than $-180°$. The phase margin is then denoted by

$$\phi_m = 180° + \arg\left[G(\epsilon^{j\omega_g T_s})G_c(\epsilon^{j\omega_g T_s})\right]$$

At the phase-crossover frequency ω_p, where $\arg[G(\epsilon^{j\omega_p T_s})G_c(\epsilon^{j\omega_p T_s})] = -180°$, the gain $G(\epsilon^{j\omega_p T_s})G_c(\epsilon^{j\omega_p T_s})$ should be less than 1 (or 0 dB). The gain margin is then denoted by

$$k_m = |G(\epsilon^{j\omega_p T_s})G_c(\epsilon^{j\omega_p T_s})|^{-1}$$

The margins k_m and ϕ_m provide information about the range of gain and phase variations in the plant that can be tolerated by the closed-loop system to maintain stability.

In addition, we can evaluate the system's performance by using the frequency response of the closed-loop transfer function from $R(z)$ to $Y(z)$ in Figure 7.8, namely

$$T(z) = \frac{G(z)G_c(z)}{1 + G(z)G_c(z)}$$

The *bandwidth* of the closed-loop system is defined as the frequency ω_B where $|T(\epsilon^{j\omega_B T_s})| = (1/\sqrt{2})|T(1)| = 0.707|T(1)|$; that is, $|T(\epsilon^{j\omega_B T_s})|$

is 3 dB below the DC gain magnitude $|T(1)|$. The bandwidth ω_B serves as an indication of the speed of a system's response to inputs. The higher the bandwidth of a system, the faster its response is.

For a system with dominant complex poles close to the unit circle such that the frequency response peak of the transfer function is greater than $|T(1)|$, we define the resonance peak as

$$M_p = \max_{\omega} |T(\epsilon^{j\omega T_s})| \tag{7.3}$$

and ω_p as the resonance frequency, where $|T(\epsilon^{j\omega_p T_s})| = M_p$.

One purpose of a feedback system is to reduce the effects of parameter variations on system performance. This reduction is not uniform over all frequencies, however. To gain further insight, we separate a gain, α, from the controller transfer function by writing $G_c(z) = \alpha \bar{G}_c(z)$. Then the change of $T(z)$ with respect to a change of α is defined as the sensitivity function

$$S(z) = \frac{\partial T(z)/T(z)}{\partial \alpha/\alpha} = \frac{1}{1 + \alpha \bar{G}_c(z) G(z)} = \frac{1}{1 + G_c(z) G(z)} \tag{7.4}$$

We can show that $S(z)$ is also the transfer function from $R(z)$ to $E(z)$ in Figure 7.8. The frequency response $S(\epsilon^{j\omega T_s})$ can be used to evaluate the robustness of the feedback system with respect to variations in the controller gain. A large value of $|S(\epsilon^{j\omega T_s})|$ indicates that the characteristics of the closed-loop system (magnitude and phase) at the frequency ω will be sensitive to the variations of $G_c(z)$.

Example 7.6 illustrates the computation of these frequency-domain performance measures. To compute the margins and the corresponding crossover frequencies, we use the Control System Toolbox `margin` function. To display the results, the `margin` function uses a Bode plot, which is valid for open-loop systems with no poles outside the unit circle.

To compute the closed-loop system bandwidth, we have created the RPI function `bwcalc`, which has three input arguments: (i) the frequency points as a column vector, (ii) the corresponding magnitude $T(\epsilon^{j\omega T_s})$ in decibels as a column vector, and (iii) the low-frequency gain $|T(1)|$ in decibels. The command `bw = bwcalc(magT_dB,w,lfgT_dB)` returns the closed-loop system bandwidth as the scalar `bw`. The variables M_p and ω_p can be obtained by applying the MATLAB `max` function to the vector `magT_dB`. In the following example, we illustrate the evaluation of these frequency-response measures.

☐ EXAMPLE 7.6 *Frequency-Response Performance*

For the sampled-data feedback system given in Example 7.1, make a Bode plot for the open-loop transfer function $G(z)G_c(z)$, and find the gain and phase margins and the related crossover frequencies of the feedback system. Then plot $|T(\epsilon^{j\omega T_s})|$ and find ω_B, M_p, and ω_p for the closed-loop system. Finally, plot $|S(\epsilon^{j\omega T_s})|$ to investigate the closed-loop system sensitivity with respect to the controller gain.

Solution The commands to compute the performance measures are contained in Script 7.6. After the discretized transfer function $G(z)$ has been obtained, the * function is used to form the open-loop transfer function $G(z)G_c(z)$. The margin function is applied to $G(z)G_c(z)$ to find the gain margin $k_m = 22$ dB and the phase margin $\phi_m = 61.6°$. The gain- and phase-crossover frequencies are 1.48 and 7.96 rad/s, respectively. A Bode plot showing these margins and the crossover frequencies is given in Figure 7.10(a).

MATLAB Script

```
% Script 7.6:  Frequency response performance
Ts = 0.1                                 % sampling time
Gp = tf(1,[1 1 0])                       % plant in continuous tf form
Gz = c2d(Gp,Ts,'zoh')                    % discretize using zero-order hold
Gc = tf(5*[1 -0.9],[1 -0.7],Ts)          % controller in discrete tf form
margin(Gz*Gc)                            % Figure 7.10(a)
T = feedback(Gz*Gc,1)                    % build CL transfer function T(z)
[mag_CLdB,ph_CL,w] = bodedb(T);          % CL gain as dB & phase
lfg_CL = dcgain(T)                       % dc gain as ratio
lfg_CLdB = 20*log10(lfg_CL)              % dc gain in decibels
bw = bwcalc(mag_CLdB,w,lfg_CLdB)         % closed-loop bandwidth in rad/s
[M_pw,ii] = max(mag_CLdB)                % maximum CL frequency response
wp = w(ii)                               % ...at this frequency (rad/s)
S = feedback(1,Gz*Gc)                    % sensitivity function S(z)
[magS_dB,phS] = bodedb(S,w);             % sensitivity frequency response
[maxS,jj] = max(magS_dB)                 % maximum sensitivity magnitude
w(jj)                                    % ...at this frequency (rad/s)
semilogx(w,mag_CLdB(:),w,magS_dB(:),'--') % Figure 7.10(b)
```

The closed-loop transfer function $T(z)$ is computed using the feedback function and is given in Example 7.1. The magnitude $|T(\epsilon^{j\omega T_s})|$ is obtained from the bodedb function in decibels in the column array magT_dB and is plotted versus ω as the solid curve in Figure 7.10(b). The low-frequency gain lfgT is computed from the dcgain function and is expressed in decibels as lfgT_dB. Together with the frequency array w, these properties are used by bwcalc to compute the closed-loop bandwidth as $\omega_B = 2.47$ rad/s. From magT_dB and w, we also obtain $M_p = 0.149$ dB and $\omega_p = 0.896$ rad/s. Note that the max command returns both the peak value in Mp and its index as the integer ii. Hence, w(ii) is the frequency at which the peak value occurred.

Treating $G(z)G_c(z)$ as the transfer function in the feedback path, the feedback function is used to generate the sensitivity function (7.4) as

$$S(z) = \frac{z^3 - 2.605z^2 + 2.238z - 0.6334}{z^3 - 2.581z^2 + 2.240z - 0.6544}$$

Its magnitude, $|S(\epsilon^{j\omega T_s})|$, is plotted versus ω as the dashed curve in Figure 7.10(b). At low frequencies, the sensitivity magnitude is small, which indicates that the low-frequency closed-loop system performance is not greatly affected by a variation in the controller gain. However, $|S(\epsilon^{j\omega T_s})|$ peaks at 2.91 rad/s with a value of 2.65 dB, indicating that the closed-loop system performance in the neighborhood of 2.91 rad/s is sensitive to any controller-gain variation. This is because the frequency of the dominant closed-loop mode is 2.31 rad/s, and this mode is strongly affected by controller gain. At high frequencies, $|T(\epsilon^{j\omega T_s})|$ approaches zero and $|S(\epsilon^{j\omega T_s})|$ approaches unity.

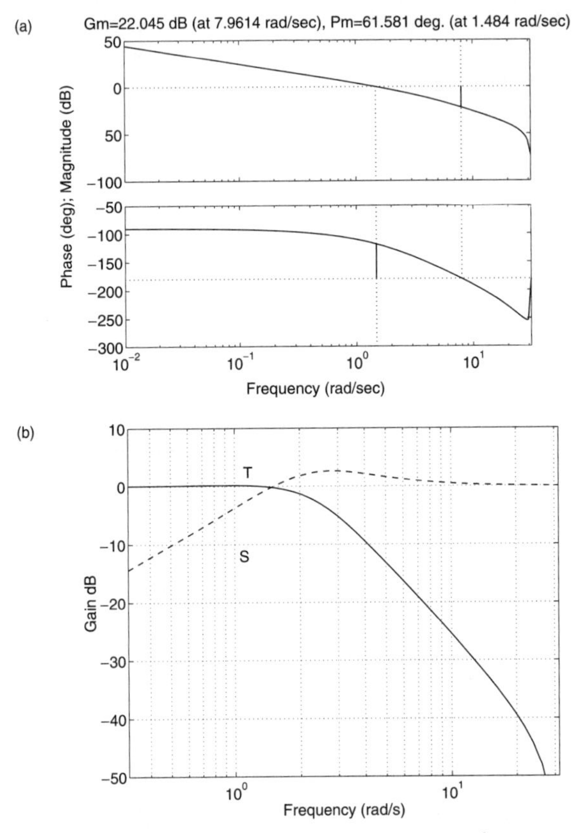

FIGURE 7.10 *(a) Bode plot indicating margins and crossover frequencies; (b) Frequency-response magnitudes of closed-loop system ($T(z)$) and sensitivity ($S(z)$) for Example 7.6*

WHAT IF? Redo Example 7.6 using the controller $G_c(z) = 100(z - 0.9)/(z - 0.7)$, which represents a gain of 10 over the controller used in the example. The increase in gain exceeds the gain margin of 16 dB = 6.31. You will find that the `margin` command responds with negative gain and phase margins, indicating that the closed-loop system is unstable. ■

□

REINFORCEMENT PROBLEMS

In each of the following problems, the plant $G_p(s)$, the controller $G_c(z)$, and the feedback-path transfer function $H(s)$ for Figure 5.10 are given. Make a Bode plot for the open-loop transfer function $\overline{HG_p}(z)G_c(z)$, and find the gain and phase margins and the related crossover frequencies. Plot the magnitude of the closed-loop frequency response $|T(\epsilon^{j\omega T_s})|$, and find the closed-loop bandwidth ω_B, the resonance peak M_p, and the resonance frequency ω_p. Also plot the magnitude of the closed-loop sensitivity function $|S(\epsilon^{j\omega T_s})|$ and comment on the closed-loop system's sensitivity with respect to the controller gain. Furthermore, find the damping ratio of the dominant poles of the closed-loop system. Simulate the unit-step response of the closed-loop system, and find the rise time t_r and the maximum overshoot M_o in percent.

P7.7 Proportional controller.

$$G_p(s) = \frac{s+5}{s^2+13s+12}, \quad G_c(z) = 9, \quad H(s) = \frac{60}{(s+3)(s+20)},$$

$$\text{and} \quad T_s = 0.05 \text{ s}.$$

P7.8 PI controller.

$$G_p(s) = \frac{1}{s^2+8s+32}, \quad G_c(z) = 20 + \frac{10z}{z-1}, \quad H(s) = 1, \text{ and } T_s = 0.1 \text{ s}.$$

NYQUIST ANALYSIS

To find the gain and phase margins of an unstable open-loop system, we apply the Nyquist technique, which requires the use of a Nyquist diagram to show the frequency response. Here we rewrite the characteristic equation $1 + G(z)G_c(z) = 0$ of the closed-loop system in Figure 7.8 as

$$G(z)G_c(z) = -1 + j0$$

The Nyquist diagram is generated by mapping $G(z)G_c(z)$ for values of z along the circumference of the unit circle in the counterclockwise direction, known as the Nyquist contour, to a closed curve $\mathcal{C}$ in a complex plane. The Control System Toolbox contains the `nyquist` function to compute and plot the curve $\mathcal{C}$. The `nyquist(G)` command directly generates the Nyquist diagram of the system `G`. The function can also be used to generate the real and imaginary parts of the frequency response, with the command `[re,im] = nyquist(G)`. Because `re` and `im` are cell arrays, we apply the `squeeze` command to remove singleton dimensions, making them into vector arrays.

The Nyquist stability criterion states that if the open-loop system has P unstable poles, the closed-loop system is stable if and only if the curve $\mathcal{C}$ encircles P times the critical point $(-1, 0)$ in the counterclockwise direction. Furthermore, from the Nyquist diagram we can find the gain margin by examining the intersections of the curve $\mathcal{C}$ with the negative real axis, and the phase margin by finding the intersection of $\mathcal{C}$ with the unit circle.

For open-loop plants having poles on the unit circle, the Nyquist contour needs to be modified to include detours around these poles. The `nyquist` command does not map the detour, and as a result, the user needs to complete the Nyquist diagram. This process can be found in many digital control textbooks and will not be covered here.

In the following example, we illustrate the use of the Nyquist diagram to determine the gain and phase margins of an unstable open-loop system.

☐ EXAMPLE 7.7 *Unstable Open-Loop System*

Use the Nyquist diagram to find the gain and phase margins of the closed-loop discrete-time system in Figure 7.8, where $T_s = 0.1$ s,

$$G(z) = \frac{1.5(z-0.5)}{z(z-1.5)} \quad \text{and} \quad G_c(z) = 1$$

Solution

In Script 7.7, the plant is entered as the LTI object `Gz`. The real and imaginary parts of the frequency response are computed from the `nyquist` command and plotted as the closed curve $\mathcal{C}$ in Figure 7.11. The upper half of the unit circle is mapped to the solid curve, and the lower part of the unit circle is mapped to the dashed curve. Note that $\mathcal{C}$ encircles the critical point $(-1, 0)$ once in the counterclockwise direction. Because the system has one unstable open-loop pole (at $z = 1.5$), $P = 1$. Based on the Nyquist stability criterion, then, the closed-loop system is stable.

To determine the gain margin, we locate the intersection of the curve $\mathcal{C}$ with the negative real axis. As shown in Figure 7.11, we have two intersections of interest, designated as `a` and `b`. Intersection `a` is at $z = -0.90$, and intersection `b` is at $z = -1.50$. These values can be found by using the zoom feature of the plot window. Thus this system has two gain margins: The lower value is $1/1.50$ $= 0.667$, and the upper value is $1/0.90 = 1.111$. The meaning of these two gain margins is that if the gain is increased beyond 1.111 or decreased below 0.667, the system will become unstable.

To find the phase margin, we locate the intersection of the solid part of $\mathcal{C}$ with the unit circle in Figure 7.11. In this case, the intersection is designated as `c` and yields the value $z = -0.32 - j0.946$. Then the phase margin ϕ_m is the angle defined by a straight line connecting `c` to the origin and the negative real axis. Thus $\phi_m = \tan^{-1}(0.946/0.32) = 71.3°$.

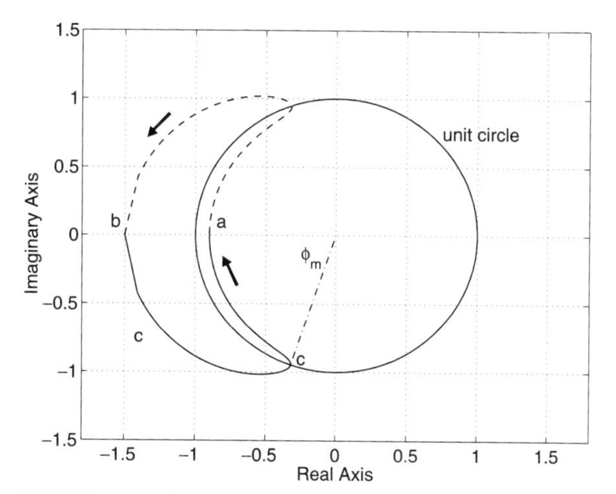

FIGURE 7.11 *Nyquist diagram indicating margins for Example 7.7*

MATLAB Script

```
% Script 7.7:  Gain and phase margins
num = 1.5*[1 -0.5];                          % plant numerator
den = conv([1 -1.5],[1 0])                    % plant denominator
Gz = tf(num,den,0.1)                          % tf object
[re,im] = nyquist(Gz);                        % generate freq response
re = squeeze(re);                             % remove singleton dimension
im = squeeze(im);                             % remove singleton dimension
plot(re,im,'-',re,-im,'--'), hold on          % make Nyquist diagram
ucircle                                       % plot unit circle
plot([0;-0.32],[0;-0.946],'-.'), hold off     % draw straight line
text(-0.85,0.1,'a')                           % label intersection points
text(-1.6,0.1,'b')
text(-0.28,-0.9,'c')
text(-0.25,-0.15,'\phi_m')                    % Greek symbol phi
```

<u>WHAT IF?</u> Use the `margin` function to assess the stability of the system considered in Example 7.7. Which of the gain margins is provided by the `margin` function in this case? ■

□

REINFORCEMENT PROBLEMS

P7.9 Proportional controller. Use the Nyquist diagram to analyze the stability and margins in Problem 7.7.

P7.10 Unstable open-loop system. Use the Nyquist diagram to analyze the stability and margins of the closed-loop system in Figure 7.8 with

$$G(z) = \frac{z^2 - z + 0.5}{z^3 - 2.1z^2 - 1.16z - 0.096}, \quad G_c(z) = \frac{z - 0.7}{z - 0.4}, \quad \text{and} \quad T_s = 0.1 \text{ s}.$$

EXPLORATORY PROBLEMS

EP7.1 Stability of audio feedback systems. Figure 7.12 shows a model of a typical public address system in an auditorium. The amplifier magnifies the microphone input by the gain K to broadcast a speech from the loudspeakers. The sound will be echoed back to the microphone with an attenuation of a and a delay of D, which are functions of the distance from the loudspeaker to the microphone and the acoustic properties of the auditorium. A lowpass filter, $H(z)$, can be added to the return path to simulate more realistically the fact that high-frequency components usually attenuate more than low-frequency components.

Model the system in Figure 7.12 in MATLAB and use the drumbeat sound track discussed in Comprehensive Problem CP5.1, which is stored in the `local.mat` file, as the input signal. Using the sampling frequency f_s = 16,384 Hz, estimate D based on the speed of sound (331 m/s) and an assumed distance between the microphone and the speaker. Define the lowpass filter to be

$$H(z) = \frac{z - 0.1}{9(z - 0.9)}$$

Assume that the attenuation is $a = 0.01$, a 40 dB reduction. Investigate the stability of the system for various values of the amplifier gain in the range $10 \leq K \leq 1000$. Simulate the loudspeaker output for each value of K selected and listen to the sound. Can you predict the value of K where the system will be unstable? For values of K slightly below the instability value, what does the loudspeaker output sound like?

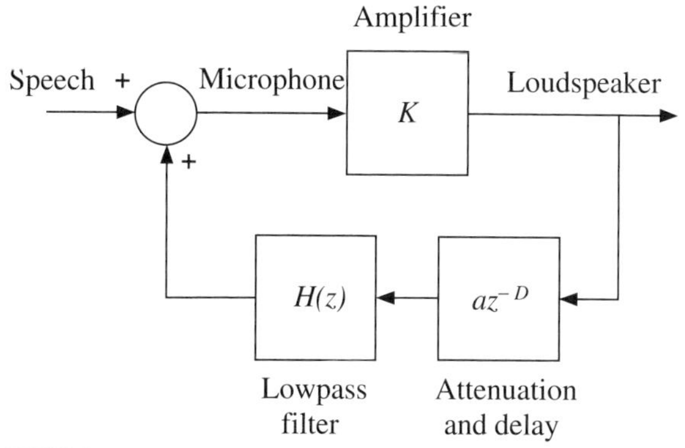

FIGURE 7.12 *Model of a public address system in an auditorium*

COMPREHENSIVE PROBLEMS

CP7.1 Ball and beam system. We use the tools discussed in this chapter to assess the performance of the controller for the ball and beam system designed in Comprehensive Problem CP3.1. Obtain the model by running the file `dbbeam.m` with the default sampling period of $T_s = 0.02$ s. Select a proportional controller, $K_m(z) = K_{P1}$, for the servo motor loop, with the gain K_{P1} to be in the range of 0.5 to 10. For each K_{P1}, find the gain and phase margins of the servo motor system. Plot the gain and phase margins as a function of K_{P1}. Apply a unit-step input to the closed-loop servo motor system and evaluate the performance measures, such as the rise time t_{rm} and the overshoot M_{om}, of the resulting step response. Plot these performance measures as a function of K_{P1}.

Based on the margins and step response, select a suitable K_{P1} to close the servo loop. Then apply a PD controller, $K_b(z)$, to control the ball position. Select a range of values from 0.4 to 2 for the proportional gain K_{P2} and from 0.4 to 2 for the derivative gain K_D. For each set of gains, find the gain and phase margins of the ball position control loop. Apply a 0.1 m step command to the closed-loop system and compute the rise time and the overshoot of the position response, and the maximum angle deviation of the wheel-angle response for each set of gains. Because the wheel angle is limited to within 90°, we need to put a constraint on the wheel-angle response. Use the `contour` function to plot the phase margin, the rise time, and the maximum wheel-angle deviation versus K_{P2} and K_D. Use the plot to find a set of PD gains to achieve the fastest rise time subject to the design requirements of the maximum wheel-angle deviation of not more than 1 rad and the phase margin of the ball position control loop of not less than 45°.

CP7.2 Electric power generation system. Consider the voltage control of the electric power system shown in Figure A.8 in Appendix A. Use a proportional-integral (PI) controller, $K_V(z)$, with its transfer function given by Equation (3.7) and vary the integral gain K_I from 0.05 to 0.5 and the gain K_P from 10 to 40. For each pair of K_I and K_P, find the gain and phase margins, and the rise time, overshoot, and settling time of the terminal voltage V_{term} subject to a 0.05 step V_{ref} input. Use the MATLAB `mesh` and `contour` functions to make 3D plots showing the dependence of these performance measures on K_I and K_P. Select a pair of gains to achieve a rise time of less than 1 s and an overshoot of less than 10%. In this design, we assume that some supplementary damping has been added to the electromechanical mode of the power system model so that the regulator design would not be affected by the stability of this mode. This is done by changing the natural damping term in the model and results in the model file `dpower2.m`, which should be used to obtain the process model for the design.

CP7.3 Hydroturbine model. Apply a proportional-integral (PI) controller as defined in Equation (3.7) for the regulation of the mechanical power produced by the turbine according to the block diagram in Figure A.10 in Appendix A, where $G(z)$ denotes the discrete-time transfer function of the series combination of the gate actuator and the hydroturbine unit. For the PI controller, try a range of gains for K_P from 0.1 to 0.4, and for K_I from 0.5 to 1.2. For each set of K_P and K_I, find the gain and phase margins, and the rise time, overshoot, and settling time of the output power response to a unit-step P_{des} input. Use the MATLAB `mesh` function to make 3D plots showing the dependence of these measures on K_P and K_I. Because the transfer function has a zero outside the unit circle, the step response initially takes a dip before turning upward. Note that the rise time is computed without taking into account this initial dip. Select a set of gains to achieve the smallest settling time, with the constraint that the phase margin be no less than 60°.

SUMMARY

In this chapter we have examined the notions of time- and frequency-domain performance measures. To complement the MATLAB functions available for computing damping ratios and gain and phase margins, we have created RPI functions for finding the velocity error constant, the step-response measures, and the closed-loop bandwidth. The Nyquist stability result for discrete-time systems was also discussed. To help you understand the significance of the z-plane poles, we included an example showing the mapping of the s-plane poles to the z-plane poles. The time-domain performance measures will be used in Chapter 8 to design PID controllers and in Chapter 10 to design state-space controllers. The frequency-domain performance measures will be used in Chapter 9 to design lead-lag controllers.

ANSWERS

P7.1 Closed-loop poles: $z = 0.3681$ and $0.8925\epsilon^{\pm j0.2119}$ – damping ratio of the complex poles = 0.4731 and natural frequency = 2.405 rad/s. Step-response performance: $M_o = 19.12\%$, $t_p = 1.40$ s, $t_r = 0.641$ s, $t_s = 3.40$ s, and $e_{ss} = 0.26\%$ (actually $e_{ss} = 0$).

P7.2 Closed-loop poles: $z = 0.9947\epsilon^{\pm j0.0315}$ – damping ratio = 0.166 and natural frequency = 0.640 rad/s; $z = 0.7033\epsilon^{\pm j0.1715}$ – damping ratio = 0.899 and natural frequency = 7.831 rad/s. Step-response performance: $M_o = 46.87\%$, $t_p = 5.35$ s, $t_r = 2.66$ s, $t_s = 31.85$ s, and $e_{ss} = 0.14$ % (actually $e_{ss} = 0$).

P7.3 *Problem P7.1 with nonideal sensor.* Closed-loop poles: $z = 0.1154, 0.3996, 0.9071\epsilon^{\pm j0.2117}$ – damping ratio = 0.409 and natural frequency = 2.39 rad/s. Step-response performance: $M_o = 25.27\%$, $t_p = 1.40$ s, $t_r = 0.60$ s, $t_s = 3.50$ s, and $e_{ss} = 0.31\%$ (actually $e_{ss} = 0$).

P7.4 *Problem P7.2 with nonideal sensor.* Closed-loop poles: $z = 0.9951\epsilon^{\pm j0.0315}$ – damping ratio = 0.153 and natural frequency = 0.637 rad/s; $z = 0.7380\epsilon^{\pm j0.2066}$ – damping ratio = 0.827 and natural frequency = 7.35 rad/s. Step-response performance: $M_o = 49.03\%$, $t_p = 5.35$ s, $t_r = 2.63$ s, $t_s = 36.50$ s, and $e_{ss} = 0.23\%$ (actually $e_{ss} = 0$).

P7.5 The system $G(z)K(z)$ is type 0 with $K_p = 2.5$ and $K_v = 0$.

P7.6 The system $G(z)K(z)$ is type 1 with $K_p = \infty$ and $K_v = 0.375$.

P7.7 Gain margin is 16.97 dB, phase margin is 70.05°, gain crossover frequency is 2.84 rad/s, and phase crossover frequency is 10.82 rad/s; closed-loop system bandwidth is 14.60 rad/s, $M_p = 2.72$ dB, and $\omega_p = 4.13$; damping ratio for poles at $z = 0.863 \pm j0.154$ is 0.597; step response: $t_r = 0.11$ s and $M_o = 45.34\%$.

P7.8 Gain margin is 12.64 dB, phase margin is 61.87°, gain crossover frequency is 3.89 rad/s, and phase crossover frequency is 10.19 rad/s; closed-loop system bandwidth is 14.64 rad/s, $M_p = 0.368$ dB, and $\omega_p = 5.18$; damping ratio for poles at $z = 0.668 \pm j0.448$ is 0.346; step response: $t_r = 0.32$ s and $M_o = 7.26\%$.

P7.9 Number of unstable open-loop poles is $P = 0$ and number of clockwise encirclements of the critical point $(-1, 0)$ by the Nyquist plot is $N = 0$; therefore the number of unstable closed-loop poles is $Z = N + P = 0$.

P7.10 Number of unstable open-loop poles is $P = 1$ and number of clockwise encirclements of the critical point $(-1, 0)$ by the Nyquist plot is $N = -1$; therefore the number of unstable closed-loop poles is $Z = N + P = 0$. Upper gain margin is 1.54, lower gain margin is 0.12; phase margin is 31.33°.

8 *Proportional-Integral-Derivative Control*

PREVIEW

Now that you have become proficient in building discrete-time system models and analyzing them with the aid of MATLAB and have established a number of measures for evaluating the performance of a feedback control system, you are ready to consider methods for designing control systems. The basic controller prototype is one that uses the error, its time derivative, and its integral with respect to time to construct the signal used to drive the actuators, which in turn affect the behavior of the process being controlled. Actually, discrete-time approximations of these variables are used, but we will refer to them as the derivative and integral of the error.

We start by showing that using a control signal proportional to the error cannot alone be expected to result in good damping and fast response and may have unacceptable steady-state error. Introducing an integral term in the controller can eliminate steady-state errors but may adversely affect the damping. A term proportional to the derivative of the error can improve the speed of response and the damping, but it does not reduce steady-state errors. A controller that combines all three terms, known as a *proportional-integral-derivative (PID) controller*, has the potential to significantly improve response time, damping, and steady-state error reduction.

PROPORTIONAL CONTROL

Consider the feedback system used in Chapter 7 to establish closed-loop system performance measures. This is shown in Figure 8.1(a). In this chapter, we consider the use of proportional, integral, and derivative functions in the controller $G_c(z)$ to meet system performance specifications. In addition to Figure 8.1(a), we will use the diagram in Figure 8.1(b) to evaluate the response of a particular design to a disturbance input.

Note that the figure does not show a sampler between the process and the sensor. This indicates that we are considering a *continuous-time* sensor with transfer function $H(s)$ that measures the continuous-time output of the process $G_p(s)$. Hence, when we need the *discrete-time* model of the process and sensor, we must use the `c2d` command on the series combination $H(s)G_p(s) = \overline{HG}_p(s)$, rather than on the individual Laplace transfer functions $H(s)$ and $G_p(s)$.

In this section we start with the design of a proportional controller with $G_c(z) = K_P$. In general, as the proportional gain K_P is increased, the rise time (t_r) and the steady-state regulation error for a unit-step reference input become smaller. These performance improvements are offset by a larger overshoot, however. Thus a typical design often involves some performance tradeoffs, and the designer must iteratively select an appropriate K_P. MATLAB is an ideal computer tool for performing this type of iterative design.

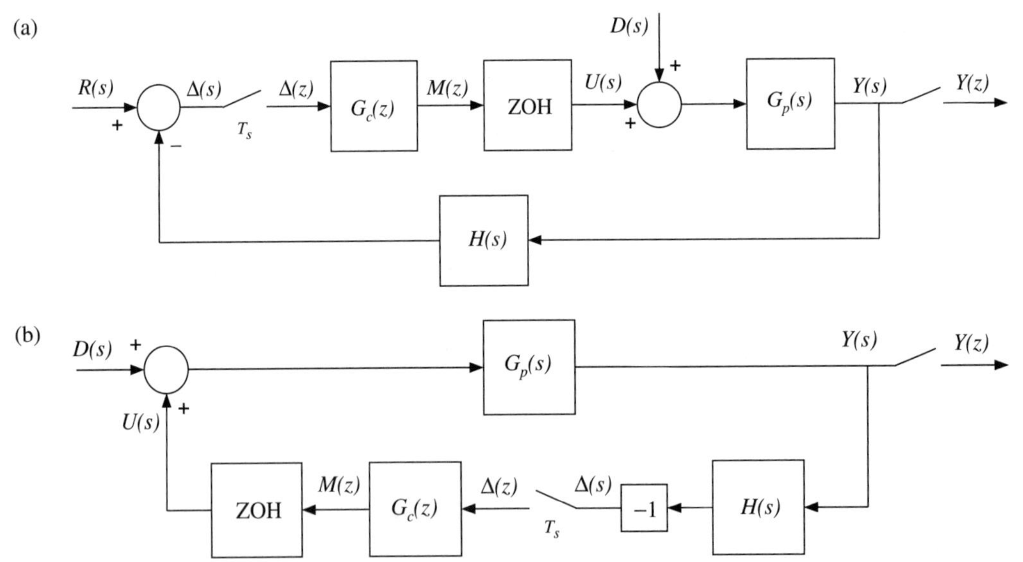

FIGURE 8.1 *Feedback system for PID controller design: (a) block diagram with reference and disturbance inputs; (b) simplified block diagram with disturbance input only*

For the selection of K_P, we generally take advantage of the computer to perform a sweep of a time simulation for a range of values for K_P and pick the gain that best satisfies the design specifications. A more traditional approach is to perform root-locus analysis, which can be readily accomplished in MATLAB. Appendix C provides an overview of the root-locus analysis technique. In Example 8.1, we illustrate the approaches of gain sweeping and root-locus analysis for controller parameter design.

Although in this chapter we will use only time-domain specifications for the design of PID controllers, we should also evaluate the final design using frequency-domain measures such as gain and phase margins and bandwidth to obtain further insight. This evaluation, when not discussed directly in the examples, will be contained in the M-files for the examples on the Brooks/Cole web site.

☐ EXAMPLE 8.1 *Proportional Control*

For the feedback control system in Figure 8.1(a) with the continuous-time process and sensor models

$$G_p(s) = \frac{4}{(2s+1)(0.5s+1)} \quad \text{and} \quad H(s) = \frac{1}{0.05s+1}$$

and a sampling time of $T_s = 0.1$ s, design a proportional controller, $G_c(z) = K_P$, following the steps given below:

a. Determine the discrete-time transfer function of the process and sensor. Then draw the root locus in the z-plane for variations in K_P and use the `rlocfind` command and the data marker feature to find K_P^*, the controller gain for which the closed-loop system becomes marginally stable. Also use the `dzline` and `rlocfind` commands to determine the value of K_P for which the system has a pair of complex closed-loop poles with a damping ratio of $\zeta = 0.8$.

b. Generate a plot that shows the responses of the closed-loop system to a unit-step reference input for several values of $K_P < K_P^*$. Use the RPI function `kstats` to find the maximum output value, rise time, peak time, steady-sate value, and percent overshoot of the step responses, and present the results in a table. Then determine, by trial and error, the largest controller-gain value $\hat{K}_P$ that results in a unit-step response having no more than 20% overshoot with respect to the steady-state output value. Determine the gain and phase margins of the feedback system with this gain.

c. Use the gain $\hat{K}_P$ to obtain a single plot showing the responses to unit-step reference and disturbance inputs, taken separately, and determine the values of the steady-state responses.

Solution

a. The MATLAB commands in Script 8.1(a) will generate the root-locus plot shown in Figure 8.2(a). The root locus has two branches starting at the open-loop poles at $z = 0.8187$ and 0.9512 that meet on the real axis at $z = 0.88$ and then move outside the unit circle as K_P is increased. Eventually, these branches come together on the negative real axis in the vicinity of $z = -5.6$, with one branch terminating on the open-loop zero at $z = -2.3096$ and the other approaching infinity along the negative real axis. A third branch starts at the open-loop pole at $z = 0.1353$ and approaches the open-loop zero at $z = -0.1426$ along the negative real axis. This asymptotic behavior is due to the fact that the open-loop system has one more pole than zeros.

When the `rlocfind` command is used with the cursor placed where the locus crosses the unit circle, we find that `kk` yields $K_P^* \approx 7$ and the three corresponding closed-loop poles are $z \approx 1.0\epsilon^{\pm j0.497}$ and 0.0886. Alternatively, the data marker capability can be used to pinpoint the unit-circle crossing, as shown in Figure 8.2(a). When the `rlocfind` command is used a second time at the point where the locus crosses the dashed line corresponding to $\zeta = 0.8$, we find that for $K_P = 0.274$ there will be a pair of complex closed-loop poles having a damping ratio of $\zeta = 0.80$. The three poles are at $z = 0.888\epsilon^{\pm j0.082}$ and 0.133.

MATLAB Script

```
% Script 8.1(a):  P-only design for 2nd-order process + 1st-order sensor
Ts = 0.1                              % sampling period
Gp_s = tf(4,conv([2 1],[0.5 1]))      % continuous process
H_s = tf(1,[0.05 1])                  % continuous sensor
HGp_s = H_s*Gp_s                      % continuous plant
HGp_z = c2d(HGp_s,Ts,'zoh')           % discrete plant, using ZOH option
rlocus(HGp_z), hold on                % draw root-locus
dzline(0.8)                           % add damping-ratio line
ucircle, hold off                     % add unit circle
axis([-1.5 1.5 -1 1])                 % get reasonable plot area
%---- find gain and CL poles where locus crosses unit circle ---
[kk1,pCL1] = rlocfind(HGp_z)          % user selects point in z plane
[mag_pCL1,theta_pCL1] = xy2p(pCL1)
%-- find gain and CL poles where locus crosses line for zeta = 0.8 --
[kk2,pCL2] = rlocfind(HGp_z)
[mag_pCL2,theta_pCL2] = xy2p(pCL2)
```

b. On the basis of the findings in part (a), we will investigate several step responses for controller gains in the interval $0 < K_P < 7$. After some experimentation, we select the range of values $K_P = 0.5, 1.0, ..., 2.5$ to illustrate the design. To produce the required plot of the step responses for a sweep of the controller gain, we use a `for` loop to cause `KP` to take on the specified values. For each gain value we build the closed-loop model by using the command `KP*Gp_z/(1+HGp_z*KP)` because the `feedback` command cannot be directly used. Then the step response is computed and plotted versus time, with the `hold` option in effect so the plots will be superimposed on a single set of axes. The `hold` option is reset after the last curve has been added to the plot.

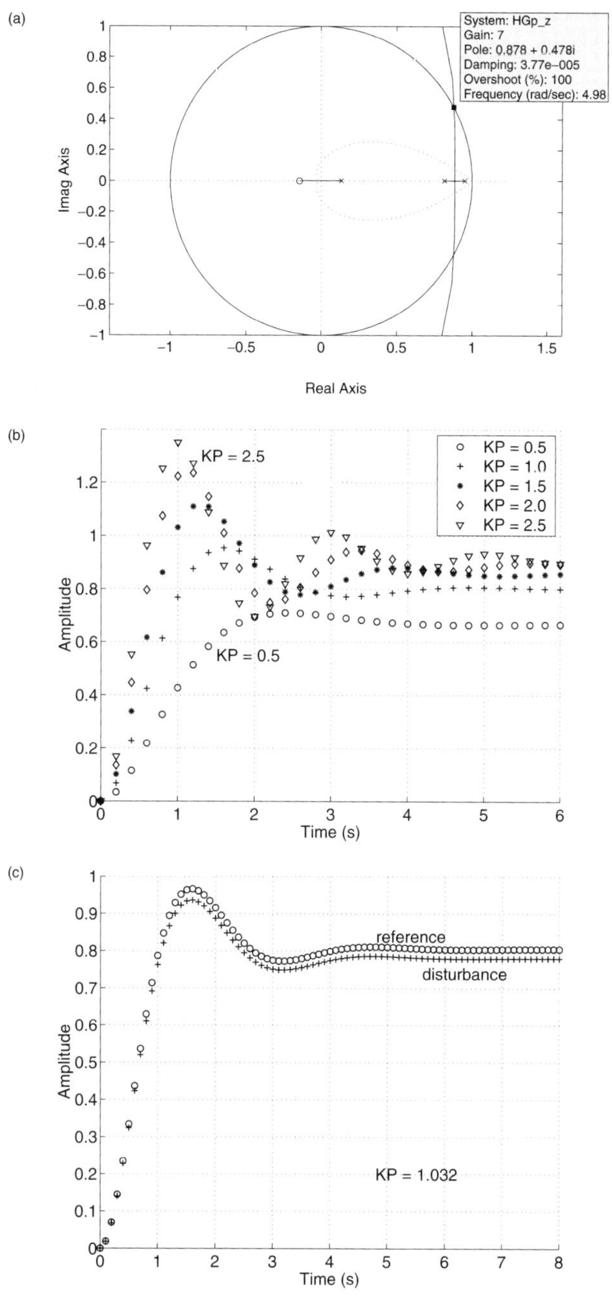

FIGURE 8.2 *Plots for proportional control: (a) root-locus plot; (b) responses to unit-step reference input for various values of* K_P*; (c) response to unit steps in the reference and disturbance inputs with* $K_P = 1.032$

MATLAB Script

```
% Script 8.1(b):  Step response with proportional gain sweep
%      Use HGp_z and Ts from Script 8.1(a)
results = [ ]                    % array to save results for building table
Pgains = [ ]                     % array for saving gains
time = 0:Ts:6;                   % vector of time points separated by Ts
figure, hold on                  % turn hold on before plotting
symbols = ['o' '+' '*' 'd' 'v']   % different plot symbols
%-- for each value of KP: build CL model, solve, and plot response
for ii = 1:5
  KP = 0.5*ii                     % KP = 0.5,1.0,...,2.5
  T = Gp_z*KP/(1 + HGp_z*KP)      % closed-loop system
  y = step(T,time);
  ref = dcgain(T);                % steady-state value of y
  [Mo,tp,tr,ts,ess] = kstats(time,y,ref);      % calc performance
  ymax = y(find(time == tp));
  yss = y(length(time));
  %-- save gains and result for making table at end ----
  results = [results; [KP  ymax  tr  tp  yss  Mo]];
  % plot every other point
  plot(time(1:2:length(time)),y(1:2:length(time)),symbols(ii))
  pause(1)                        % wait 1 second before doing next gain
end
legend('KP = 0.5','KP = 1.0','KP = 1.5','KP = 2.0','KP = 2.5')
hold off                          % turn hold off after last curve
```

Figure 8.2(b) shows the results of the gain sweep. For the lower values of K_P there is a longer rise time, less overshoot, and a larger steady-state error. For the higher values of K_P there is a shorter rise time, more overshoot, and a smaller steady-state error. Clearly, our options for improving the performance of the closed-loop system are rather limited with the proportional controller.

Within the `for` loop of the step-response generation, the percent overshoot M_o is computed using the `kstats` function with the step response `y` and the time array `time` as the first two input arguments. Here the overshoot is defined with respect to the steady-state value of the output y_{ss}, which becomes the third input argument to `kstats`. Because the transients have mostly decayed at the end of the simulation, y_{ss} is taken to be the last element in `y`. The time needed to reach the peak value (t_p) and the rise time (t_r) are also computed. Table 8.1 shows the numerical results obtained for the range of controller gains specified in Script 8.1(b). By adjusting the values of `KP` in the `for` command to be confined to a smaller interval, or by entering specific values of `KP` manually and executing the instructions contained within the `for` loop, we find that $K_P = 1.032$ will result in an overshoot of 20%.

K_P	$y_{\max}$	t_r	t_p	y_{ss}	M_o (%)
0.5000	0.7089	1.1713	2.4000	0.6659	6.3354
1.0000	0.9540	0.7166	1.6000	0.7996	19.2495
1.5000	1.1178	0.5431	1.3000	0.8573	30.4141
2.0000	1.2445	0.4498	1.1000	0.8936	40.0013
2.5000	1.3508	0.3903	1.0000	0.8973	48.5934

TABLE 8.1 *Performance measures for the step responses generated by the gain sweep in Example 8.1*

Applying the command `[Gm,Pm,Wcg,Wcp] = margin(KP*HGp_z)` with $K_P = 1.032$, we obtain a gain margin of $k_m = 6.622$, which is equivalent to 20 $\log_{10}$ $(6.622) = 16.42$ dB. This result indicates that the controller gain that will result in a marginally stable closed-loop system is $K_P^* = 1.032 \times 10^{16.42/20} = 6.83$. This number is close to the value of 7 obtained with the `rlocfind` command. Also note that the value of the phase-crossover frequency ω_p, at which the gain margin is measured, is 4.92 rad/s. From the root-locus plot in Figure 8.2(a), we see that the upper-right root-locus branch crosses the unit circle at an angle of 0.497 rad. We know that the frequency values along the unit circle are given by $z = \epsilon^{j\omega T_s}$, where T_s is the sampling period. Because the sampling period is 0.1 s, we have $\omega = 0.497/0.1 = 4.97$ rad/s, which agrees quite well with the value of ω_p obtained from the `margin` command.

c. With $K_P = 1.032$, construct the closed-loop system model shown in Figure 8.1(a) and simulate the step response with the `step` command, without plotting it. These operations are implemented in the first half of Script 8.1(c) and result in the column vector `y_ref`.

Then we build the closed-loop model shown in Figure 8.1(b), for which the input is $D(z)$, the sign at the feedback summing junction is positive, and the numerator of the feedback transfer function is negative. Thus the closed-loop system command `Gp_z/(1+HGp_z*KP)` requires a plus sign in the denominator. The `step` command produces the column vector `y_dist`. Next we plot the two step responses versus time in Figure 8.2(c). Finally, the steady-state response values are obtained from the last two statements in the script file; they are $(y_{\text{ref}})_{\text{ss}} = 0.8052$ and $(y_{\text{dist}})_{\text{ss}} = 0.7803$. The performance of the proportional control is not entirely satisfactory, because of the large steady-state error and the impact of the disturbance on the system response.

MATLAB Script

```
% Script 8.1(c):  Closed-loop reference & disturbance step responses
T_ref = KP*Gp_z/(1 + HGp_z*KP)      % build CL system with reference input
t = 0:Ts:8;                         % define time vector for plots
y_ref = step(T_ref,time);           % CL step response to reference input
T_dist = Gp_z/(1 + HGp_z*KP)        % build CL system with disturbance input
y_dist = step(T_dist,time);         % CL step response to disturbance input
plot(time,y_ref,'o',time,y_dist,'+')  % plot both responses
y_ref_ss = y_ref(length(time))      % final value for reference input
y_dist_ss = y_dist(length(time))    % final value for disturbance input
```

WHAT IF?

a. Suppose the design specification in part (b) of Example 8.1 is to achieve a rise time of no more than 1 s. Find the smallest value of K_P that will satisfy the specification.

b. Suppose the design specification in part (b) of Example 8.1 is to achieve $y_{ss} \geq 0.9$. Find the smallest value of K_P that will satisfy the specification. ■

□

PROPORTIONAL-INTEGRAL CONTROL

In Example 8.1 we showed that proportional control can yield well-damped responses to both reference and disturbance step inputs, but steady-state errors should be expected. Although increasing the controller gain will reduce the steady-state errors, it will also increase the overshoot and reduce the damping. Clearly, a more sophisticated controller is needed to achieve significantly reduced or zero steady-state errors with acceptable overshoot and damping. In this section we pursue these objectives by introducing an integral term in the control law.

The form of the proportional-integral (PI) controller we will use is

$$m(k) = K_P \left[\Delta(k) + K_I T_s \sum_{\ell=0}^{k} \Delta(\ell) \right] \tag{8.1}$$

where $\Delta(k)$ is the controller input and $m(k)$ is the controller output, as shown in Figure 8.1(a), and ℓ is a dummy variable of summation. The z-transform of the summation operation is approximated by

$$\mathcal{Z}\left\{ \sum_{\ell=0}^{k} \Delta(\ell) \right\} \approx \left(\frac{z}{z-1} \right) \Delta(z)$$

Thus the transfer function of the controller is $G_c(z) = M(z)/\Delta(z)$, which can be written as

$$G_c(z) = K_P \left[1 + K_I T_s \left(\frac{z}{z-1}\right)\right] \tag{8.2}$$

showing clearly the contributions of the individual terms. To make obvious the locations of the zeros and poles of the PI controller, we rewrite the transfer function as a rational function, namely

$$G_c(z) = K_P \left[\frac{(1 + K_I T_s)z - 1}{z - 1}\right] \tag{8.3}$$

which agrees with the transfer function in Equation (3.7). In this form we see that $G_c(z)$ has a pole at $z = 1$ and a zero at $z = 1/(1 + K_I T_s)$, which will lie to the left of the pole. Using this controller, we will be able to ensure that a stable feedback control system will achieve zero steady-state errors for step inputs.

Our approach will be to use MATLAB root-locus analysis and simulation in the same manner as in Example 8.1, with sweeps of the controller gain K_P for several values of the integral gain K_I. On the basis of the results, we will select values for K_I and K_P to satisfy the design specifications.

☐ EXAMPLE 8.2 *PI Control*

For the system in Example 8.1, design a PI control law according to Equation (8.2) by performing the following steps:

a. Plot the reference unit-step responses for a sweep of the proportional gain K_P from 0.3 to 1.1 in increments of 0.2 with integral-gain values of $K_I = 0, 0.3, 0.4, 0.5, 0.6$, and 1.0. Select a value, $\hat{K}_I$, that will result in good damping and a 2% settling time $t_s \leq 5$ s.

b. Plot the root locus for variations in K_P with $K_I = \hat{K}_I$. Then use the `rlocfind` command to calibrate several points on the locus in terms of K_P. Find K_P^*, the gain for which the closed-loop system becomes marginally stable.

c. With $K_I = \hat{K}_I$, select the largest K_P such that the overshoot for a reference unit-step input is less than 10%. Then compute and plot the unit-step responses to both reference and disturbance inputs. Comment on the differences between these responses and those in Figure 8.2(c) obtained with a proportional controller.

Solution

a. Script 8.2(a) contains the commands to accomplish the building of the closed-loop model and the generation of the step-response plots. First, we form the plant transfer function $G_p(s)$ and the feedback transfer function $H(s)$ as LTI objects in TF form and discretize the cascade connection $H(s)G_p(s) = \overline{HG}_p(s)$. Because the proportional gain `KP` is being varied inside the `for` loop, the controller transfer function $G_c(z)$ must be computed inside the loop, before applying the closed-loop system command.

The response curves for $K_I = 0$ will be the same as those obtained with proportional control and shown in Figure 8.2(b), because $G_c(z)$ reduces to K_P in this case. The step responses for $K_I = 0.3, 0.5$, and 1.0 are shown in Figure 8.3. The responses in Figure 8.3(a) for a low integral gain ($K_I = 0.3$) appear to be headed toward $y = 1.0$, but the transients are dying out slowly, resulting in a large settling time t_s. Increasing K_I to 0.5, shown in Figure 8.3(b), causes the transients to be reduced to virtually zero for all $t > 6$ s, for all values of K_P used. For a high integral gain ($K_I = 1.0$), Figure 8.3(c) shows that the responses no longer have sufficient damping, which results in a longer settling time. From this analysis, we conclude that the responses for $K_I = 0.5$ will satisfy the constraint of 2% settling time $t_s \leq 5$ s. Accurate values of t_s can be found using the `kstats` command.

MATLAB Script

```
% Script 8.2(a):  Sweep of KP for selected values of KI
Ts = 0.1                                    % sampling period
Gp_s = tf(4,conv([2 1],[0.5 1]))            % continuous process
H_s = tf(1,[0.05 1])                        % continuous sensor
HGp_s = H_s*Gp_s                            % continuous plant
HGp_z = c2d(HGp_s,Ts,'zoh')                 % discrete plant
%----- compute CL step responses for range of KP values
time  = 0:Ts:8;
KI = input('enter integral gain KI ==> ') % user specifies KI
figure, hold on
symbols = ['o' '+' '*' 'd' 'v']             % different plot symbols
for ii = 1:5
  KP = 0.3 + 0.2*(ii-1)                     % KP = 0.3,0.5,...,1.1
  Gc  = KP*tf([1+KI*Ts  -1],[1 -1],Ts);     % PI control law
  T = Gp_z*Gc/(1 + HGp_z*Gc)                % closed-loop system
  y = step(T,time);                         % closed-loop step response
  ref = dcgain(T);
  [Mo,tp,tr,ts,ess] = kstats(time,y,ref)  % response stats
  % plot every other point
  plot(time(1:2:length(time)),y(1:2:length(time)),symbols(ii))
end
legend('KP = 0.3','KP = 0.5','KP = 0.7','KP = 0.9','KP = 1.1')
hold off
```

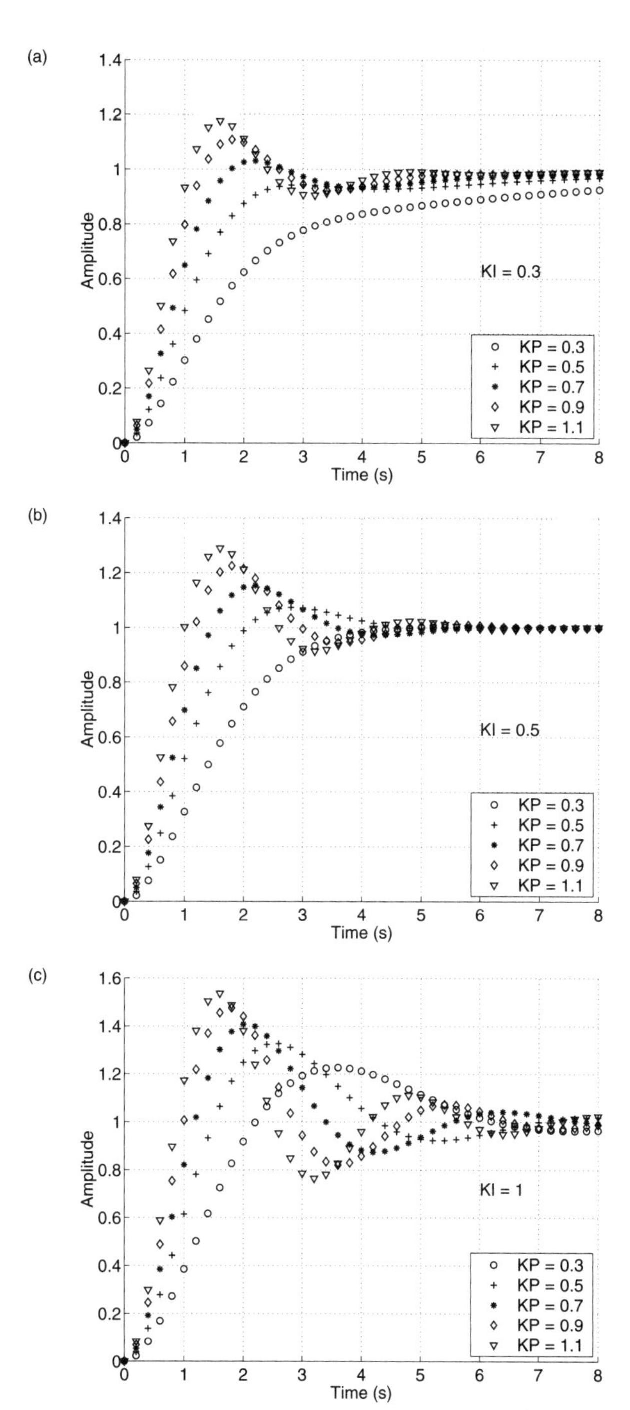

FIGURE 8.3 *Results of proportional-gain sweep with PI controller: (a)* $K_I = 0.3$*; (b)* $K_I = 0.5$*; (c)* $K_I = 1.0$

b. Using Script 8.2(b) for $K_I = 0.5$, the required root locus is plotted in Figure 8.4. Note that the controller transfer function $G_c(z)$ has a pole at $z = 1$ and a zero at $z = 1/1.05 = 0.9524$, resulting in the controller zero in effect canceling the plant pole at $z = 0.9512$.

By repeated use of the `rlocfind` command, we determine the values of K_P that correspond to the point at which two branches of the locus intersect ($z = 0.91$) and to five other points on the upper complex branch of the locus shown in Figure 8.4. From the locus we see that for $K_P = 0.216$, the closed-loop system will have a double pole at $z = 0.91$. For $K_P = 5.28$, there will be a complex pole where the locus crosses the unit circle, at $z \approx 1.0\epsilon^{\pm j0.430}$. The numerical values of the poles can be displayed in the MATLAB command window by using the `rlocfind` command, where `kk` contains the required proportional gain and the `polesCL` array gives the corresponding closed-loop poles.

To see the effect of the change in going from proportional control to PI control, compare the root-locus plots in Figures 8.2(a) and 8.4. Observe that the net effect of the integral term with $K_I = 0.5$ is to replace the open-loop plant pole at $z = 0.9512$ with a new pole at $z = 1$. This change shifts the complex portion of the locus for the PI controller slightly to the right, making the complex branches cross the unit circle at a lower value of ω than for proportional control (about 4.30 rad/s versus the value of 4.97 rad/s found in part (a) of this example). Also, the point at which the large-gain asymptotes intersect shifts slightly to the right ($\sigma_0 = 0.886$ for proportional control versus 0.910 for PI control). There is also a third branch of the locus, starting at the open-loop pole of $\overline{HG_p}(z)$ at $z = 0.1353$, going to the left along the real axis, and terminating at the plant zero at $z = -0.1426$. The transients associated with the closed-loop pole on this branch will decay much faster than those associated with the other poles (because the pole is closer to $z = 0$ than the poles on the complex branches). Thus, this real pole has very little effect on the response, which is primarily caused by the two complex closed-loop poles.

MATLAB Script

```
% Script 8.2(b):  Root-locus plot for PI control
% Use HGp_z from Script 8.2(a)
KI = 0.5
KP = 1                                % fix KP at unity for RL plot
Gc  = KP*tf([1+KI*Ts  -1],[1 -1],Ts)  % controller
Gol = HGp_z*Gc
rlocus(Gol)
hold on, ucircle, hold off            % add unit circle plot
axis([0.6 1.12 -0.1 0.55])            % adjust region plotted
[kk,polesCL] = rlocfind(Gol)          % calibrate locus with gains
```

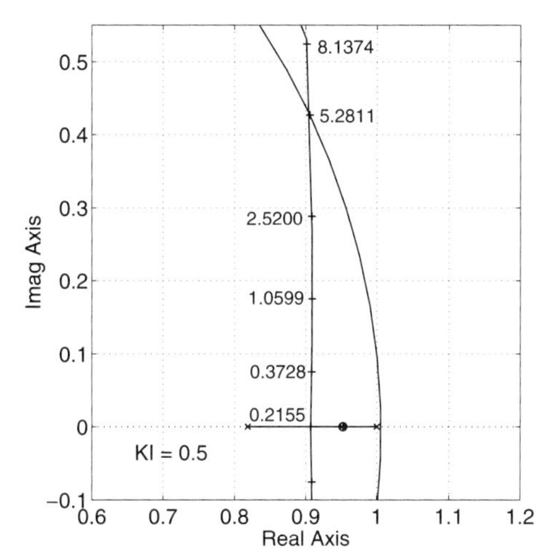

FIGURE 8.4 *Root-locus plot for PI control with gain calibrations added*

c. On the basis of the responses shown in Figure 8.3(b), we can estimate that $K_I = 0.5$ and $0.5 \leq K_P \leq 0.7$ will result in approximately 10% overshoot for a step input. By adjusting K_P with $K_I = 0.5$ by trial-and-error, we find that $K_P = 0.563$ is the largest proportional gain for which the overshoot does not exceed 10%. The corresponding settling time is 4.0 s, which more than satisfies our design specification.

The reference and disturbance step responses for these gains can be computed with a M-file almost identical to Script 8.1(c). The only changes required are: (i) the controller transfer function has the numerator `KP*[1+KI*Ts -1]` and the denominator `[1 -1]`, and (ii) the user must be allowed to specify the integral gain `KI` in addition to `KP`. The results, shown in Figure 8.5, illustrate that the steady-state errors for both inputs go to zero. For the reference step response, there is a 10% overshoot and a 2% settling time of 4.0 s, satisfying the design specifications. The response to the disturbance input reaches a somewhat higher value than with the proportional controller (1.150 versus 0.936), but it settles to 0.052 in 8 seconds and has a steady-state value of zero, whereas with proportional control the steady-state error is 0.780. Thus we have graphic evidence, at least for the process and sensor models used, of the ability of the PI controller to eliminate steady-state errors to both reference and disturbance step inputs. Essentially, the integral term in the PI controller has increased the system type from 0 to 1. It is well known that a type-1 system will exhibit zero steady-state error to a step input.

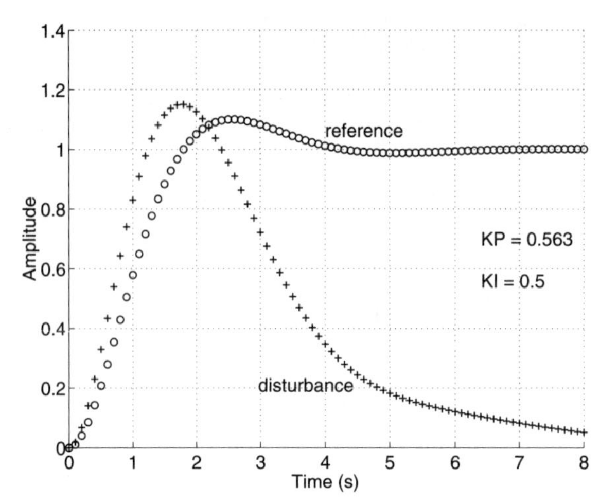

FIGURE 8.5 *Reference and disturbance step responses with PI controller gains of* $K_P = 0.563$ *and* $K_I = 0.5$

WHAT IF?

a. If $K_I = 0.3$ or 1.0 is used in the PI controller, there will no longer be an approximate pole-zero cancellation. Draw root-locus plots for these values of K_I and examine closely the change in the behavior of the root locus near the point $z = 1$. Relate your observations to the step responses shown in Figures 8.3(a) and (c).

b. Suppose the reference and disturbance inputs are ramp functions. What sort of steady-state errors would you expect? Copy the file `ex8_2c.m` for this example and modify it to produce the responses of this closed-loop system to ramp inputs with unity slopes, taken one at a time. ■

□

PROPORTIONAL-INTEGRAL-DERIVATIVE CONTROL

With a PI controller, we would expect a control system to satisfy design specifications of overshoot, settling time, and zero steady-state error. If a faster rise time, t_r, without increased overshoot is desired, however, then a derivative term may need to be included in the controller. With the discrete-time approximation of the derivative term added to the PI control law in (8.1), the proportional-integral-derivative (PID) control law is

$$m(k) = K_P \left[\Delta(k) + K_I T_s \sum_{\ell=0}^{k} \Delta(\ell) + K_D \left(\frac{\Delta(k) - \Delta(k-1)}{T_s} \right) \right] \quad (8.4)$$

where $\Delta(k)$ is the controller input and $m(k)$ is the controller output, as shown in Figure 8.1(a), and ℓ is a dummy variable of summation. The transfer function of the PID control law is $M(z)/\Delta(z)$; this can be written as

$$G_c(z) = K_P \left[1 + K_I \left(\frac{T_s z}{z-1} \right) + K_D \left(\frac{z-1}{T_s z} \right) \right] \quad (8.5)$$

To gain some insight about the zeros and poles of the PID controller's transfer function, we rewrite (8.4) as a rational function of z, namely

$$G_c(z) = K_P \left[\frac{(1 + K_I T_s + K_D/T_s)z^2 - (1 + 2K_D/T_s)z + K_D/T_s}{z(z-1)} \right] \quad (8.6)$$

In this form we see that $G_c(z)$ still has the pole at $z = 1$, as the PI controller transfer function has, but now there is an additional pole at $z = 0$, and two zeros that can be real or complex, depending on the values the designer selects for the gains K_I and K_D.

We now extend the designs of Examples 8.1 and 8.2 by incorporating a derivative term in the controller. The same gain-sweep approach using simulation and root-locus analysis will be applied. The gain sweep for a PID design will be more involved, however, as there are more interactions among the proportional, derivative, and integral control actions. In general, the integral gain K_I can still be set as if designing a PI controller. However, the proportional and derivative gains K_P and K_D need to be tuned together with two performance measures—namely, overshoot and rise time—to be satisfied. In addition, it is likely that there will be several combinations of K_P and K_D that can satisfy the specifications.

☐ EXAMPLE 8.3 *PID Control*

For the feedback control system considered in Examples 8.1 and 8.2, design a PID controller such that the unit-step reference response satisfies the specifications of (i) overshoot $\leq 10\%$, (ii) rise time $t_r \leq 0.35$ s, and (iii) 2% settling time $t_s \leq 1.4$ s, by following the gain-sweep procedure outlined below.

In the PI design of Example 8.2, we determined that the integral gain $K_I = 0.5$ will satisfy $t_s \leq 6$ s. As a result, we fix K_I at that value and adjust K_P and K_D using the following steps:

a. With $K_I = 0.5$, plot the reference unit-step responses for a sweep of the proportional gain K_P from 1.7 to 2.3 in increments of 0.2 for derivative-gain values of K_D =0.3, 0.4, 0.5, and 0.6. On the basis of the results, select a second set of K_P and K_D values to be searched with K_D increments of 0.02 and K_P increments of 0.1.

b. Run a series of gain sweeps with this finer mesh and use the results to select a pair of gain values $\hat{K}_P$ and $\hat{K}_D$ that produces a reference step response that performs well for all three of the specifications.

c. Plot the root locus for variations of K_P with $K_I = 0.5$ and $K_D = \hat{K}_D$. Then use the `rlocfind` command iteratively to locate the points on the locus corresponding to $K_P = \hat{K}_P$ and determine all of the closed-loop poles. Also, determine the values of K_P for several points along the branch that moves upward from the real axis and eventually passes through the unit circle.

d. With $K_I = 0.5$, $K_D = \hat{K}_D$, and $K_P = \hat{K}_P$, compute and plot the unit-step responses to both reference and disturbance inputs over the interval $0 \le t \le 8$ s. Compare these responses with those in Figure 8.2(c) obtained with a proportional controller and those in Figure 8.5 obtained with a PI controller.

Solution

The M-files required for the PID controller design can be readily obtained by starting with the files that solve Example 8.2 and modifying the expression for `Gc` to incorporate the derivative term according to (8.5). Also, an outer `for` loop is required to handle the iteration in K_D. These files are not displayed here, but they can be obtained from the Brooks/Cole web site.

K_P	K_D	t_r	t_s	M_o (%)	
1.7000	0.3000	0.4362	2.0000	7.4937	
1.9000	0.3000	0.3836	1.8000	8.8632	
2.1000	0.3000	0.3354	1.6000	10.5655	
2.3000	0.3000	0.3074	1.4000	12.7997	
1.7000	0.4000	0.3912	0.6000	1.6174	
1.9000	0.4000	0.3291	0.9000	3.1784	**
2.1000	0.4000	0.2921	0.9000	6.3906	**
2.3000	0.4000	0.2548	0.8000	10.1981	
1.7000	0.5000	0.3293	1.3000	0.8145	**
1.9000	0.5000	0.2803	1.2000	3.0380	**
2.1000	0.5000	0.2438	1.2000	7.0207	**
2.3000	0.5000	0.2240	1.1000	12.4420	
1.7000	0.6000	0.2774	1.5000	0	
1.9000	0.6000	0.2386	1.4000	5.9360	**
2.1000	0.6000	0.2159	1.2000	12.0213	
2.3000	0.6000	0.1954	1.1000	17.2156	

TABLE 8.2 *Performance measures for the step responses generated by the first gain sweep in Example 8.3*

a. Rather than show plots of the step responses, we include Table 8.2, which gives the numerical values of the three specifications (rise time, 2% settling time, and overshoot percentage). These values were obtained using the RPI function `kstats`.

From the table, we can see that six cases satisfy all three of the specifications. These have the symbol ** appended to the corresponding row in the table. Based on these results, we restrict our attention for the second set of gain sweeps to the region defined by $0.4 \le K_D \le 0.6$ and $1.9 \le K_P < 2.1$. It should be noted that for the case where $K_P = 1.70$ and $K_D = 0.30$, the value of 2.0000 for t_s signifies that the response had not settled to within $\pm 2\%$ of the steady-state value of unity by the end of the simulation (2.0 seconds in this case). If the simulation were run for a longer time, the responses would have settled and `kstats` would have returned the corresponding results.

b. Running the same MATLAB code as for the first set of gain sweeps, but with the limits and increments of the two `for` loops modified accordingly, we obtain the results shown in Table 8.3. Examining the results, we see that all but the last two combinations of K_P and K_D satisfy all the specifications. We observe that the case for which $K_P = 1.95$ and $K_D = 0.45$ has the following characteristics: (i) one of the lowest overshoot percentages (3.27%), (ii) a rise time (0.296 s) that is well below the specification, and (iii) is tied for the lowest

K_P	K_D	t_r	t_s	M_o (%)
1.9000	0.4000	0.3291	0.9000	3.1784
1.9500	0.4000	0.3197	0.9000	4.0858
2.0000	0.4000	0.3105	0.9000	4.9259
2.0500	0.4000	0.3013	0.9000	5.6937
2.1000	0.4000	0.2921	0.9000	6.3906
1.9000	0.4500	0.3068	0.7000	2.3482
1.9500	0.4500	0.2962	0.7000	3.2728
2.0000	0.4500	0.2854	1.1000	4.4765
2.0500	0.4500	0.2746	1.1000	5.6171
2.1000	0.4500	0.2635	1.1000	6.6950
1.9000	0.5000	0.2803	1.2000	3.0380
1.9500	0.5000	0.2674	1.2000	4.1399
2.0000	0.5000	0.2547	1.2000	5.1705
2.0500	0.5000	0.2492	1.2000	6.1305
2.1000	0.5000	0.2438	1.2000	7.0207
1.9000	0.5500	0.2522	1.3000	3.5687
1.9500	0.5500	0.2463	1.3000	4.7594
2.0000	0.5500	0.2406	1.2000	6.3823
2.0500	0.5500	0.2350	1.2000	7.9555
2.1000	0.5500	0.2297	1.2000	9.4790
1.9000	0.6000	0.2386	1.4000	5.9360
1.9500	0.6000	0.2327	1.3000	7.5419
2.0000	0.6000	0.2269	1.3000	9.0912
2.0500	0.6000	0.2214	1.2000	10.5843
2.1000	0.6000	0.2159	1.2000	12.0213

TABLE 8.3 *Performance measures for the step responses generated by the second gain sweep in Example 8.3*

settling time (0.70 s). Hence, rather than selecting one combination at random, we select this set of three gains for plotting the reference and disturbance responses.

c. Taking $K_I = 0.50$ and $K_D = 0.45$, we construct the root-locus plot for K_P shown in Figure 8.6. The points on the branch of the locus that moves upward from the real axis indicates the closed-loop poles corresponding to the values of K_P shown beside them. Plus symbols indicate the corresponding closed-loop poles on the other branches. For the proportional gain $K_P = 1.9595$, which is very close to the design value, the closed-loop poles are $z = 0.9104\epsilon^{\pm j0.0099}, 0.6410\epsilon^{\pm j0.5895}$, and -0.0689. The dominant closed-loop poles for this value of the proportional gain are close to $z = 0.53 \pm j0.36$, which correspond to a damping ratio of $\zeta = 0.6$.

d. Using all three design gains—namely, $K_P = 1.95$, $K_I = 0.50$, and $K_D = 0.45$—we generate and plot the closed-loop responses to step-function reference and disturbance inputs. The results are shown in Figure 8.7. A comparison of the reference-input responses for the three types of controls shows a marked improvement for the response with PID control over both of the others. The response with proportional control is deficient in both speed of response and steady-state error. The response with PI control has zero steady-state error, but its rise time is much slower than that with PID control.

Likewise, the disturbance-input response obtained with PID control shows a peak of about 0.361 and zero steady-state error, which is a significant improvement over the other two controls. The disturbance response with proportional control [see Figure 8.2(c)] has a peak of 0.936 and a steady-state error of 0.780. With PI control, the disturbance response decays to zero in the steady state,

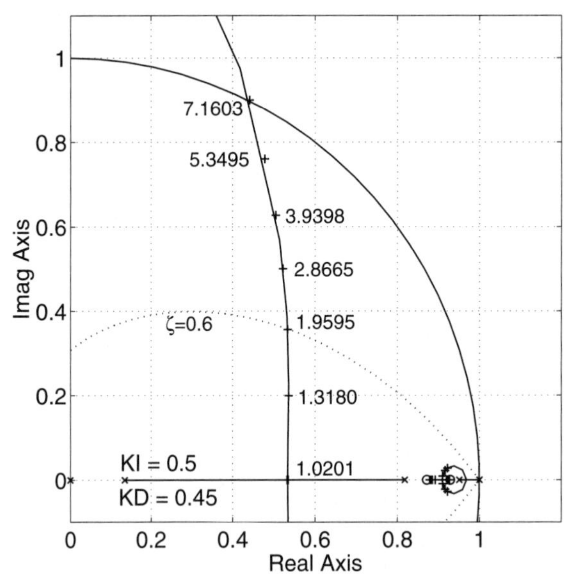

FIGURE 8.6 *Root-locus plot of PID control with gain calibration added*

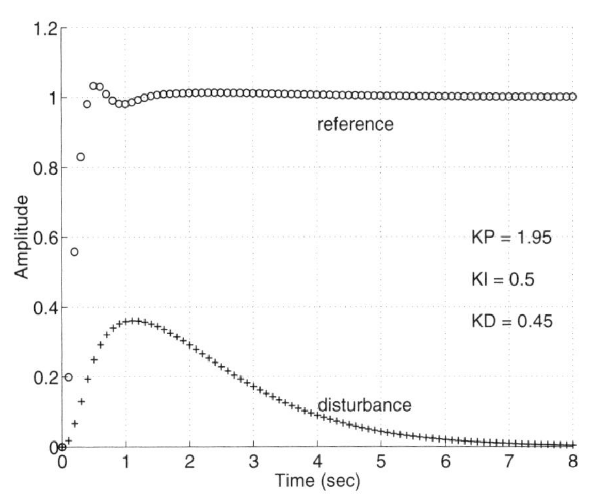

FIGURE 8.7 *Reference and disturbance step responses for PID control with final controller gains*

but from Figure 8.5 we see that its peak value is 1.150. Clearly the PID controller provides greatly improved performance over the other two, at very little increase in complexity.

WHAT IF?

a. Plot the root-locus in Example 8.3 (PID controller) for variations in the proportional gain K_P, with $K_I = 0.50$ and $K_D = 0.45$, which correspond to the system with the step responses shown in Figure 8.7 when $K_P = 1.95$. Examine the shape of the loci near the point $z = 1$.

b. Investigate the PID controller for values of K_I other than 0.5. See if you can find any combinations of the three controller gains that yield improvements in the responses, such as faster rise and settling times, while keeping the overshoot percentage between 5 and 10% for a step in the reference input.

c. Design a proportional-derivative (PD) controller for the feedback system considered in these examples. The control law is that of the PID controller (8.5), with the integral gain K_I set to zero. You should be able to obtain a rapid response, but there will be steady-state errors comparable to those obtained with proportional control.

d. Suppose the reference and disturbance inputs are ramp functions. What sort of steady-state errors would you expect? Copy the file `ex8_3d.m` for Example 8.3 and modify it to obtain the responses of this closed-loop system to ramp inputs with unity slopes, taken one at a time. ■

□

EXPLORATORY PROBLEMS

EP8.1 Type-0 plant. Consider the feedback control system in Figure 8.1(a) for the type-0 fifth-order plant and the sensor whose transfer functions are given by

$$G_p(s) = \frac{500(s+10)}{(s+2)(s+5)(s^2+6s+25)(s+20)} \quad \text{and} \quad H(s) = 1$$

where the sampling period is selected to be $T_s = 0.05$ s. Design a PID controller whose resulting reference step response satisfies as many of the following specifications as possible (hopefully all of them):
a. no more than 10% overshoot
b. 10 to 90% rise time no more than 0.6 s
c. time-to-peak no more than 1.5 s
d. 2% settling time no more than 2.8 s
e. zero steady-state error

Your MATLAB program should use the RPI function `kstats` to compute these performance measures for the step response of your final design. Follow the steps described in Examples 8.1 to 8.3 and use root-locus plots and appropriate gain sweeps to help obtain a satisfactory set of controller gains.

EP8.2 Type-1 plant. A type-1 plant, $G_p(s)$, and a sensor, $H(s)$, in the feedback system of Figure 8.1(a) have the transfer functions

$$G_p(s) = \frac{12,000}{s(s+10)(s+30)(s+40)} \quad \text{and} \quad H(s) = \frac{1}{0.02s+1}$$

The sampling period is $T_s = 0.005$ s, which is one quarter of the sensor time constant. Design a PD controller that provides a step response with no more than 10% overshoot and a 2% settling time of no more than 1 s. Verify that the steady-state error to the step input is zero but that there is a nonzero steady-state error to a ramp input. Calculate the velocity constant of your design and relate it to the steady-state error.

COMPREHENSIVE PROBLEMS

CP8.1 Ball and beam system. Use a combined root-locus and gain-sweep approach to design $K_m(z)$ and $K_b(z)$ for the ball and beam control system in Figure A.3 in Appendix A. The inner-loop involving the controller $K_m(z)$ should be designed first, followed by the design for the controller $K_b(z)$. The model for the ball and beam system discretized at 50 Hz can be obtained from the `dbbeam.m` file.

a. Proportional-controller design: Obtain the transfer function from V_{in} to the wheel angle θ and design a proportional controller $K_m(z) = K_{P1}$ for the servo-motor to control the wheel angle θ. Perform a root-locus analysis of this control loop and select the gain K_{P1} that yields a damping ratio of 0.9.
b. PD-controller design: With the wheel-angle control loop closed using $K_m(z) = K_{P1}$ found from part a, develop the model from θ_{des} to the ball position ξ. Design a proportional-derivative controller according to Equation (3.6) for $K_b(z)$ to control the ball position. In this design, select a range of values K_D between 0.8 and 2. For each value of K_D, perform a root-locus analysis with respect to the proportional gain K_{P2}. In the root-locus plot, there is a pair of dominant poles, one of which starts at $z = 1$, moving toward each other and eventually becoming a complex conjugate pair. Select the value of K_{P2} that achieves a 0.5 damping ratio for these dominant modes. Simulate the response to a step command of the closed-loop system for each set of PD gains and find the rise time and the maximum wheel angle deviation. Select a set of PD gains that achieves a rise time of not more than 3 s, with the wheel angle not exceeding 1 rad.

CP8.2 Inverted pendulum. Run the `dstick.m` file to obtain the model of the inverted pendulum discretized with a sampling frequency of 100 Hz. Comprehensive Problem CP3.2 shows that the pendulum cannot be stabilized by a PD feedback controller applied to the angle signal because the transfer function from the motor voltage input to the pendulum angle has a real zero at $z_0 = 1$ and a real pole outside the unit circle (at $z_p = 1.06$). In this problem, you will design the controllers $K_p(z)$ and $K_c(z)$ shown in Figure A.6 of Appendix A to stabilize the pendulum and move the cart to a desired location.

Start by designing an unstable first-order pendulum controller, $K_p(z)$, by locating the controller pole between $z_0 = 1$ and $z_p = 1.06$. Locate the controller zero between $z_0 = 1$ and $z = 0.95$. Perform a root-locus analysis and obtain a stabilizing design with a damping ratio of 0.8 for the closed-loop poles near $z = 1$. Next, with the $K_p(z)$ loop closed, design a proportional controller for $K_c(z) = K_P$ by performing a root-locus analysis to minimize the rise time for a step response. This can be done by selecting a K_P from the root-locus plot and performing a simulation for a position command of 0.1 m. You should observe that this design will not be able to achieve a rise time of much better than 10 s, which is quite slow. A state-space design can achieve a much faster response, with a rise time of less than 2 s, which is the design objective of Comprehensive Problem CP10.2.

CP8.3 Electric power system. Use the gain-sweep approach to design the controllers outlined below. In this design, use the model file `dpower2.m`, which has increased damping on the electromechanical mode of the power system model. Use a sampling frequency of 100 Hz, as given in `dpower2.m`. The terminal bus voltage variable is used as the output for both designs.

a. P-controller design: Design a proportional controller with the transfer function $K_V(z) = K_P$ for the voltage control of the electric power system shown in Figure A.8 of Appendix A. For a step reference input of magnitude 0.05, the steady-state error should be less than 5% and the 2% settling time should not exceed 5 s.
b. PI-controller design: Design a proportional-integral controller with transfer function given by Equation (8.2), for the voltage controller $K_V(z)$ of the electric power system shown in Figure A.8 of Appendix A. For a step reference input of magnitude 0.05, the rise time should be no more than 1 s and the 2% settling time no more than 5 s.

CP8.4 Hydroturbine model. The discretized hydroturbine system has a zero outside the unit circle, which causes the step response to go briefly negative before going positive. Consider designing for the system a PID controller whose transfer function is given in Equation (8.5), with the integral part to achieve zero steady-state error to step commands and the derivative part to reduce the time-delay effect. For a unit-step power command P_{des}, the design objectives are to have a damping ratio of at least 0.4, a rise time of no more than 1 s, and a settling time of no more than 14 s.

SUMMARY

In this chapter we have discussed the design of PID controllers to satisfy the time-domain performance measures discussed in Chapter 7. Because there are no exact relationships between the PID gain parameters and the time-domain performance measures for high-order systems, a gain-sweep approach based on time simulation was used to select the controller gains to satisfy the performance measures. We also used root-locus plots to improve our understanding of the effects of the controller gains on the step responses. In Chapter 9 we consider the design of lead-lag controllers using frequency-response methods and as an approximation to a PID design.

9 Frequency-Response Design

PREVIEW

In the previous chapter, we presented several examples in which proportional, proportional-integral (PI), and proportional-integral-derivative (PID) compensators were designed. Other useful compensators include lead and lag controllers, which will be addressed in this chapter. For continuous-time systems, the parameters for lead-lag compensation can be readily derived based on the frequency response. Discrete-time frequency response offers fewer direct clues on selecting the poles and zeros of the compensator to achieve certain design requirements, such as gain and phase margins. On the other hand, a bilinear transform can be applied to a discrete-time model to derive an equivalent continuous-time system. Then continuous-time lead-lag design techniques can be applied to the resulting system to obtain the desired continuous controller. A digital controller can then be obtained by exercising the inverse bilinear transform.

Consider the sampled-data feedback control system shown in Figure 9.1(a), where $R(s)$ is the Laplace transform of the input command, $G_p(s)$ is the transfer function of the continuous plant, $G_{h0}(s) = (1 - \epsilon^{-T_s s})/s$ is the transfer function of the zero-order hold, $G_c(z)$ is the transfer function of the digital controller, and T_s is the sampling period. Assuming ideal sampling, the equivalent discrete-time feedback control system is shown in Figure 9.1(b), where

$$G(z) = \mathcal{Z}\{G_p(s)G_{h0}(s)\} \tag{9.1}$$

For this discussion, we do not represent the sensor dynamics $H(s)$ in the feedback loop. This is done to keep the block diagrams in Figure 9.1 as simple as possible because $H(s)$ cannot be separated from $G_p(s)$ in the z-transform (9.1).

In Chapter 8, after $G(z)$ had been obtained, we used gain sweeps and root-locus plots to directly design the digital controller $G_c(z)$. This approach is useful when damping ratios and step-response rise times, overshoots, and settling times are the main design objectives. When design objectives include robustness-related specifications such as phase and gain margins, however, it is desirable to consider a design in the frequency domain. In the s-domain, lead-lag compensation can be designed based on properties such as the dB-per-decade magnitude rolloff and the degree-per-decade change in phase. As discussed in Chapter 6, such properties

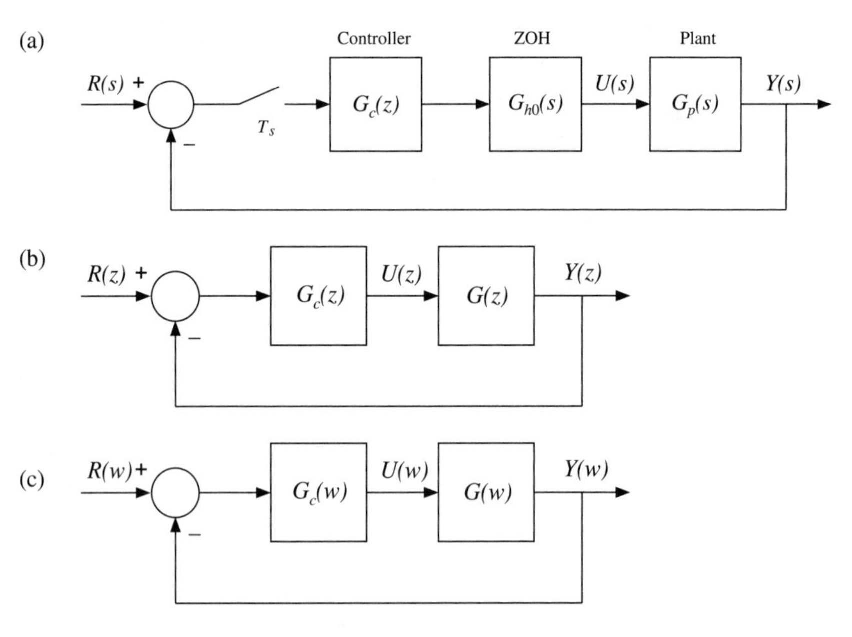

FIGURE 9.1 *(a) Closed-loop sampled-data system; (b) discretized system; (c) continuous-time equivalent in the w-domain*

are much less transparent in the z-domain. For this reason, it is desirable to transform the discrete-time system $G(z)$ to an equivalent continuous-time system and perform the control design in the continuous-time domain.

In Chapter 6 several transformations were proposed for obtaining equivalent discrete-time systems from a continuous-time system. The inverses of these transformations can be used for obtaining continuous-time equivalent systems from a discrete-time system. Among these transformations, only the bilinear transform pair, given by

$$w = \frac{2}{T_s}\left(\frac{z-1}{z+1}\right) \quad \text{and} \quad z = \frac{1+(T_s/2)w}{1-(T_s/2)w} \tag{9.2}$$

preserve the frequency response and the stability property of the closed-loop system. By this statement we mean that if the closed-loop w-domain system is stable, the closed-loop z-domain system is guaranteed to be stable. Note that in (9.2) we use w as the Laplace variable to distinguish it from the s variable for the continuous plant $G_p(s)$.

An important property of the bilinear transform is that it maps the imaginary axis in the w-plane onto the unit circle in the z-plane. For any frequency ω in the z-domain, there corresponds a unique frequency in the w-domain given by

$$\omega_w = \frac{2}{jT_s}\left(\frac{\epsilon^{j\omega T_s}-1}{\epsilon^{j\omega T_s}+1}\right) = \frac{2}{T_s}\tan\frac{\omega T_s}{2} \tag{9.3}$$

It follows that this mapping is approximately linear for small ω, but becomes nonlinear at higher frequencies. In particular, $w = j\omega_w$ for ω_w large will map into the vicinity of $z = -1$. From (9.3), the frequency responses at the two frequencies ω and ω_w are identical—that is,

$$G(w)|_{w=j\omega_w} = G(z)|_{z=\epsilon^{j\omega T_s}} \tag{9.4}$$

Consider Figure 9.1(c), in which the continuous plant $G(w)$ is obtained from the bilinear transformation as

$$G(w) = G(z)|_{z=\frac{1+(T_s/2)w}{1-(T_s/2)w}} \tag{9.5}$$

Suppose that the continuous-time controller $G_c(w)$ is designed so that the closed-loop system is stable in the w-domain and achieves a gain margin k_m and a phase margin ϕ_m. If the discrete-time controller $G_c(z)$ is obtained from $G_c(w)$ as

$$G_c(z) = G_c(w)|_{w=\frac{2}{T_s}\frac{z-1}{z+1}} \tag{9.6}$$

and applied to the closed-loop system in Figure 9.1(b), then from (9.4) the closed-loop discrete-time system will achieve the same gain margin k_m and phase margin ϕ_m. Thus the bilinear transform preserves the frequency response properties of the feedback control system.

In the following example, we plot the frequency responses of $G(z) = G(\epsilon^{j\omega T_s})$ and $G(w) = G(j\omega_w)$ and show their correspondence. The example system will be used for control design in the subsequent examples of this chapter.

☐ EXAMPLE 9.1 *Frequency Response Plots of $G(z)$ and $G(w)$*

For the continuous-plant transfer function $G_p(s)$ in Figure 9.1(a) given by

$$G_p(s) = \frac{4}{(2s+1)(0.5s+1)} \tag{9.7}$$

find $G(z)$ and $G(w)$ when the sampling time is $T_s = 0.1$ s. Plot the frequency responses of $G(z) = G(\epsilon^{j\omega T_s})$ and $G(w) = G(j\omega_w)$. For $\omega = 5\pi$ rad/s, use the bilinear transform (9.3) to compute the corresponding frequency ω_w and determine the values of $G(z)$ and $G(w)$ at $z = \epsilon^{j\omega T_s}$ and $\omega = j\omega_w$, respectively.

Solution

In Script 9.1 we begin by entering the transfer function $G_p(s)$ and performing the discretization using `c2d` with the `zoh` option. The result is

$$G(z) = \frac{0.01842z + 0.001695}{z^2 - 1.770z + 0.7788} \tag{9.8}$$

The discrete-to-continuous transformation `d2c` with the `tustin` option applied to $G(z)$ yields

$$G(w) = \frac{-0.0004149w^2 - 0.1910w + 3.986}{w^2 + 2.493w + 0.9965} \tag{9.9}$$

Note that $G(w)$ is displayed in MATLAB in terms of "`s`" because of the `d2c` command, but it really is "`w`". The frequency-response plots of $G(z)$ and $G(w)$ are shown in Figures 9.2(a) and (b), respectively. These two frequency responses are identical except for the fact that the frequency scales are different, due to the nonlinear effect of the bilinear transformation on the frequencies. For this sampling rate of 10 Hz, however, the frequency responses of both systems look alike up to 10 rad/s, because the bilinear transform at low frequencies is mostly linear.

As a final verification, we compute the frequency ω_w in the w-domain corresponding to the frequency $\omega = 5\pi$ rad/s in the z-domain as

$$\omega_w = \frac{2}{jT_s}\left(\frac{\epsilon^{j5\pi T_s} - 1}{\epsilon^{j5\pi T_s} + 1}\right) = 20 \tag{9.10}$$

Then the real and imaginary parts of $G(z)$ and $G(w)$ at those frequencies are computed using the `bode` command as

$$G(z)|_{z=\epsilon^{j5\pi T_s}} = G(w)|_{w=j20} = 0.0140\epsilon^{j2.522} \tag{9.11}$$

MATLAB Script

```
% Script 9.1 Frequency Response Plots
num = 4                              % numerator
den = conv([2 1],[0.5 1])            % denominator
Gp = tf(num,den)                     % continuous lti object
Ts = 0.1                             % sampling period
Gz = c2d(Gp,Ts,'zoh')                % discretization
Gw = d2c(Gz,'tustin')                % bilinear transform
bode(Gz)                             % continuous-time frequency response
bode(Gw)                             % discrete-time frequency response
w = 5*pi
ww = 2/Ts*tan(w*Ts/2)                % Eqn (9.3)
[magz,phz] = bode(Gz,w)              % find magnitude and
[magw,phw] = bode(Gw,ww)             % ... phase
```

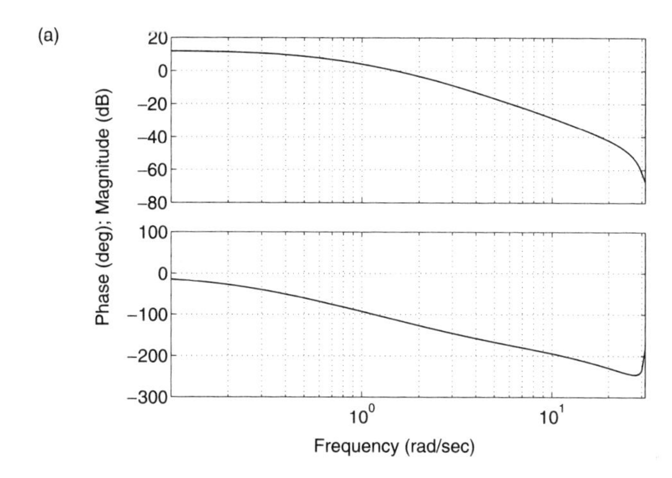

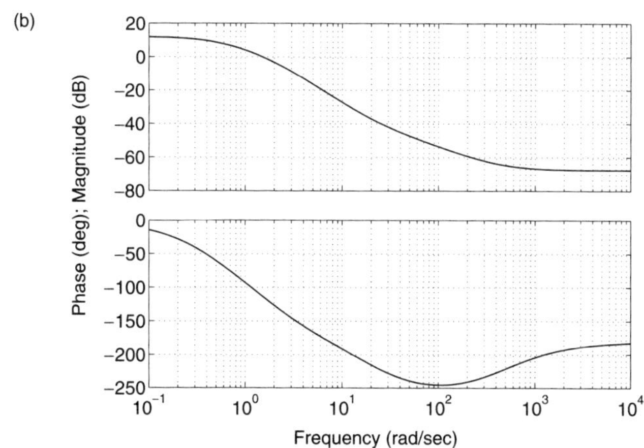

FIGURE 9.2 *Frequency Response of (a) $G(z)$ and (b) $G(w)$ for Example 9.1*

□

LEAD AND LAG CONTROLLERS

Lead and lag controllers are alternatives to PID controllers that can be considered for control design. The lead controller is similar to a PD controller; it is useful for improving the response speed and the settling time. The lag controller is similar to a PI controller; it is useful for reducing the step-response steady-state tracking error and overshoot. The lead-lag controller, formed by the cascade connection of a lead and a lag controller, has characteristics similar to a PID controller.

Consider a first-order continuous-time transfer function in the w-domain in the form

$$G_c(w) = K_w \left(\frac{w - w_z}{w - w_p} \right) \tag{9.12}$$

with $K_w > 0$, the pole at $w = w_p < 0$, and the zero at $w = w_z < 0$. Equation (9.12) is the transfer function of a lead controller if w_z is closer to the origin than w_p—that is, $w_p < w_z < 0$—resulting in a frequency response with positive (lead) phase. Otherwise, if $w_z < w_p < 0$, it represents a lag controller having a frequency response with negative (lag) phase.

Using the bilinear transform (9.6), $G_c(w)$ can be converted to an equivalent digital controller

$$G_c(z) = K_z \left(\frac{z - z_0}{z - z_p} \right) \tag{9.13}$$

where the zero is

$$z_0 = \frac{2 + w_z T_s}{2 - w_z T_s}$$

the pole is

$$z_p = \frac{2 + w_p T_s}{2 - w_p T_s}$$

and the gain is

$$K_z = K_w \left(\frac{2 - w_z T_s}{2 - w_p T_s} \right)$$

From (9.4) we know that the bilinear transform preserves the controller's lead and lag characteristics. A lead controller $G(w)$ will be transformed into a lead controller $G_c(z)$ having its zero closer to the point $z = 1$ than the pole. Otherwise, $G_c(z)$ is a lag controller. The lead and lag controllers correspond, respectively, to the highpass and lowpass filters discussed in Chapter 6.

In the remainder of this chapter, we discuss the use of frequency-response methods for designing lag controllers, lead controllers, and lead-lag controllers (obtained by a series connection of lead and lag controllers).

Once $G(w)$ has been obtained, we can apply frequency-domain design techniques to obtain controllers to meet the desired performance specifications. In this section, we discuss a lag-compensator design method, followed by a lead-compensator design method in the next section.

When designing a lag compensator for feedback systems having the structure shown in Figure 9.3, which now includes the sensor dynamics model in the feedback loop, we seek to attain at least a specified phase margin (for dynamic response) and low-frequency gain (for steady-state error to a step input). When using frequency-response methods, there are two ways in which the design can be approached. One approach is to achieve a satisfactory phase margin and use the lag to increase the low-frequency gain. The other approach is to start with the gain that meets the steady-state error specification and then to use the lag to reduce the mid- and high-frequency magnitude so the phase-margin requirement can be meet. In this section we will demonstrate only the first approach. Readers interested in the second approach are referred to the companion continuous-time MATLAB problem book by Frederick and Chow (2000).

Franklin, Powell, and Emami-Naeini (1994) present a method for designing lag compensators that attempts to satisfy phase-margin and low-frequency–gain specifications and starts at the high end of the frequency spectrum. A paraphrase of their lag compensation design procedure follows. The controller transfer function is written as

$$G_c(w) = K_{\text{lag}} \left(\frac{w - z_{\text{lag}}}{w - z_{\text{lag}}/\alpha_{\text{lag}}} \right) \tag{9.14}$$

where $\alpha_{\text{lag}} > 1$. Given the desired values for the phase margin and for the low-frequency gain of the open-loop system (to ensure a desired steady-state error to a constant reference input), we must determine values for the gain K_{lag}, for the zero-pole ratio α_{lag}, and for the controller zero z_{lag}. The magnitude of z_{lag} is also known as the *corner frequency* of the zero. The steps that will accomplish this are:

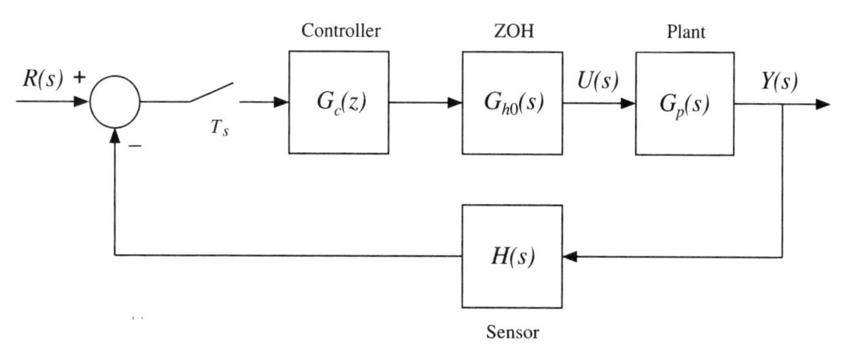

FIGURE 9.3 *Feedback system for controller design*

1. With a proportional controller, we find the gain K_{lag} such that the phase-margin specification is satisfied, without being concerned with the low-frequency gain. Having done this, we designate the gain-crossover frequency as ω_{wc} and denote the low-frequency gain attained with this controller as A. To be on the safe side, we adjust the value of K_{lag} to obtain a bit more than the required phase margin (say, 5° to 10° more).

2. Next we compute the value of α_{lag} as the ratio of the desired low-frequency gain to A, the gain attained with the proportional-only control in the previous step.

3. Finally, we select the corner frequency of the lag to be between one decade and one octave below ω_{wc}, the magnitude-crossover frequency found in step 1. Hence, we look for $\omega_{wc}/10 \leq -z_{\text{lag}} \leq \omega_{wc}/2$. The compensator pole will be $p_{\text{lag}} = z_{\text{lag}}/\alpha_{\text{lag}}$. Then we draw a Bode plot of the open-loop system with the lag controller in series with the plant and the sensor, or we use the `margin` command to determine the actual phase margin. If necessary, we repeat this process for different values of z_{lag} until we are able to meet the phase-margin specification. At this point the initial design is complete, and we plot the response of the system to the reference step input to assess the behavior in the time domain.

Depending on the step response, we may want to tune on one or more of the three design parameters to attain a more satisfactory response. For example, the reference step response may have more overshoot than desired, or it may have a slowly decaying transient. To deal with these situations, we can adjust K_{lag} to increase the phase margin or increase the magnitude of z_{lag} to make the transient decay faster.

In Example 9.2, which deals with the same system as Example 8.1 did, we apply these steps to the design of a lag controller for the plant and sensor. The system is type-0, so the steady-state error is $e_{\text{ss}} = 1/(1+K_p)$, where K_p is the position constant, given by Equation (7.1). Because we will not be able to attain zero steady-state error to a step-reference input, we will settle for a specification of 2% steady-state error. This means that the low-frequency gain of the open-loop system must be at least $(1/0.02) - 1 = 49$. Also, when we made a Bode plot of the open-loop frequency response of the final design in Example 8.1, we found that its phase margin was 60.4°. In the lag design, we will reduce the phase margin requirement to not less than 50°, so that the time response will not be too slow.

☐ EXAMPLE 9.2 *Lag Controller Design*

Using the frequency-response design algorithm described above, find a lag compensator that results in a phase margin of approximately 50° and an open-loop low-frequency gain of 49 for the system in Figure 9.3, where the transfer functions of the plant and sensor are given by

$$G_p(s) = \frac{4}{(2s+1)(0.5s+1)} \quad \text{and} \quad H(s) = \frac{1}{0.05s+1} \tag{9.15}$$

respectively. Use a sampling period of $T_s = 0.1$ s.

Solution

We present the design in two parts. In the first part, we determine values of K_{lag} and α_{lag} that provide the required low-frequency gain and meet the phase-margin specification when the lag zero is selected. In the second part, we try one or more values for the lag zero and see what we get for the phase margin.

a. The commands in Script 9.2(a) build a series interconnection of the open-loop plant $G_p(s)$ and the sensor $H(s)$. Then $\overline{GH}(z) = \mathcal{Z}\{H(s)G_p(s)G_{h0}(s)\}$ is computed using the `c2d` function, where $G_{h0}(s)$ is the zero-order hold transfer function, yielding

$$\overline{GH}(z) = \frac{0.0080859(z+2.310)(z+0.1426)}{(z-0.9512)(z-0.8187)(z-0.1353)}$$

MATLAB Script

```
% Script 9.2(a): Determination of Klag, alpha, and wwc
tauP1 = 2, tauP2 = 0.5, tau_sen = 0.05, Kproc = 4 % parameters
Gp = tf(Kproc,conv([tauP1 1],[tauP2 1]))          % plant Gp(s)
H  = tf(1,[tau_sen 1])                            % sensor H(s)
GH = H*Gp          % OL transfer function is plant * sensor in s-domain
Ts = 0.1                                          % sampling period
GHz = c2d(GH,Ts,'zoh')                            % discretization
GHw = d2c(GHz,'tustin')                           % bilinear transform
%-------------- open-loop frequency response
ww = logspace(-1,1,100)';           % use 100 points for better resolution
[mag_db,ph] = bodedb(GHw,ww);       % computes mag & phase
mag_ratio = 10.^(mag_db/20);        % convert from db to ratio
%-------- display table of index, phase, mag, & freq values
%           by selecting -135 <= phase <= -115 deg
for ii = find((ph <= -115) & (ph >= -135))
  disp([ ii  ph(ii) mag_ratio(ii)  ww(ii)])
end
%--- interpolate to get magnitude for 55 deg phase margin
mag125 = interp1(ph,mag_ratio,-125)
Klag = 1/mag125                         % set gain for selected phase
lfg = Klag*dcgain(GHw)                  % lfg for plant + gain
Alag = 49/lfg             % lag alpha supplies rest of req'd low-freq gain
wwc = interp1(ph,ww,-125)
bode(GHw)                               % plot open-loop frequency response
```

Then we compute the bilinear transformation

$$\overline{GH}(w) = \overline{GH}(z)\big|_{z=\frac{1+(T_s/2)w}{1-(T_s/2)w}} \tag{9.16}$$

using the `d2c` function to obtain

$$\overline{GH}(w) = \frac{0.002254(w-50.54)(w-20)(w+26.65)}{(w+15.23)(w+1.993)(w+0.4999)}$$

The magnitude and phase of the open-loop frequency response $\overline{GH}(w)|_{w=j\omega_w}$ are computed at 100 logarithmically spaced frequencies in the interval $0.1 \leq \omega_w \leq 10$ rad/s. Here we use the RPI function **`bodedb`** so that the output magnitude and phase are column vectors.

Allowing a 5° safety margin, we find the gain K_{lag} that results in $50+5 = 55°$ of phase margin. This can be done by using MATLAB to search the phase-angle results of the **`bodedb`** command and find the index of the entry for which the phase angle is closest to $55 - 180 = -125°$. Then we set K_{lag} equal to the reciprocal of the corresponding magnitude value. To assist in this process, we use the **`find`** command, as shown in the script, to extract the index, phase angle, magnitude, and frequency of those entries having phase angles in the interval $[-135°, -115°]$. The result is shown in Table 9.1, from which we can see that the index `ii` $= 62$ corresponds to a phase of $-123.8°$, a magnitude of 0.8537, and a frequency of 1.71 rad/s. Thus, by making $K_{\text{lag}} = 1/0.8537 = 1.084$, we obtain the desired phase margin, and the gain-crossover frequency is $\omega_{wc} = 1.71$ rad/s.

A more direct way of finding the magnitude corresponding to a phase angle of $-125°$ is to use MATLAB's **`interp1`** command to do a linear interpolation with the vectors **`mag_ratio`** and **`ph`**. If we think of the magnitude as a function of the phase angle, we then seek the magnitude value that corresponds to the phase angle $-125°$. Rather than building Table 9.1 and doing an approximate interpolation visually, we can have MATLAB obtain K_{lag} by entering the command **`mag125 = interp1(ph,mag_ratio,-125)`** followed by **`Klag = 1/mag125`**. The result is $K_{\text{lag}} = 1/0.8284 = 1.207$. To get an accurate value for the frequency ω_{wc} at which the phase angle is $-125°$, we can reuse the command **`interp1`** as **`wwc = interp1(ph,ww,-125)`**. The result of this calculation is $\omega_{wc} = 1.747$ rad/s.

Index	*Phase (deg)*	*Magnitude*	*Frequency (rad/s)*
59	−116.4	1.023	1.49
60	−118.9	0.965	1.56
61	−121.3	0.908	1.63
62	−123.8	0.854	1.71
63	−126.3	0.801	1.79
64	−128.8	0.751	1.87
65	−131.2	0.703	1.96
66	−133.7	0.658	2.06

TABLE 9.1 *Selected frequency-response values for the plant and sensor in Example 9.2*

The magnitude and phase plots for the plant and sensor alone are shown in Figure 9.4, from which we can see the graphical significance of these calculations. Horizontal lines have been added to the phase plot at $-125°$ and to the magnitude plot at $20\log_{10} 0.8284 = -1.635$ dB, and a vertical line has been drawn at the frequency 1.747 rad/s.

By using the `dcgain` command for the plant-sensor combination and multiplying the result by the value of K_{lag}, we find that the open-loop low-frequency gain is 4.828. We can then calculate the value of the zero/pole ratio as $\alpha_{\text{lag}} = 49/4.828 = 10.15$, which results in an overall low-frequency gain of 49.

b. To implement the lag compensator $G_c(w)$, we use the commands in Script 9.2(b), with $K_{\text{lag}} = 1.207$ and $\alpha_{\text{lag}} = 10.15$. The value of the lag zero is entered by the user via the `input` command. It is anticipated that the magnitude of the zero should be about 10% less than 1/10 of $\omega_{wc} = 1.747$ rad/s. After trying several values in the recommended interval $[\omega_{wc}/15, \omega_{wc}/5]$, we determine that a value of $z_{\text{lag}} = -0.157$ results in a phase margin of 50.24°, which satisfies the design specification of 50°. Thus we have designed the continuous-time controller

$$G_c(w) = \frac{1.207(w + 0.157)}{w + 0.01547} \tag{9.17}$$

which, by using (9.2) with $T_s = 0.1$ s, is equivalent to the digital controller

$$G_c(z) = \frac{1.216(z - 0.9844)}{z - 0.9985} \tag{9.18}$$

Note that because the pole and zero of $G_c(z)$ are very close to the unit circle at $z = 1$, it is important that they be represented accurately in the controller. In MATLAB, when such a situation arises, it is a good practice to convert the system representation to the state-space form rather than the polynomial form.

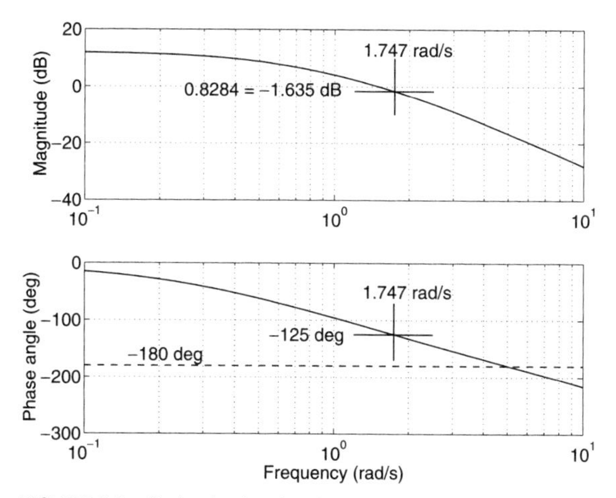

FIGURE 9.4 *Bode plot for the plant and sensor in Example 9.2*

MATLAB Script

```
% Script 9.2(b):  select lag zero & calc phase margin
%     Note:  variables defined in Script 9.2(a) are in workspace
w_Zlag = input('Enter corner freq for lag zero .....  ')
Zlag = -w_Zlag                              % lag zero must be negative
Gcw = tf(Klag*[1 -Zlag],[1 -Zlag/Alag])     % lag controller
GcGHw = Gcw*GHw          % connect lag in series with plant and sensor
lfg = dcgain(GcGHw)                         % low-freq gain of current design
[km,pm,wkm,wpm] = margin(GcGHw);            % open-loop frequency response
disp([20*log10(km)  wkm])                   % gain margin in dB & omega_gm
disp([pm wpm])                  % phase margin in degrees & omega_pm
ess = 1/(1 + dcgain(GcGHw))     % verify that steady-state error is OK
Gcz = c2d(Gcw,Ts,'tustin')      % controller in z domain
% reference response
Gz = c2d(Gp,Ts,'zoh')           % discretize plant
Klagz = ss(Klagz)               % convert to ss form for better numerics
Tz = Gz*Klagz/(1+GHz*Klagz)     % closed-loop transfer function
step(Tz,8)                      % step response to 8 s
```

Figure 9.5 shows the open-loop compensated system magnitude and phase angle plots for the initial stage (dashed) and for the final design (solid). An examination of the solid curves verifies that the phase-margin and low-frequency-gain specifications have been satisfied. The figure also clearly illustrates the effect of the pole and the zero of the lag compensator on the magnitude and phase angle of the frequency response. At low frequencies the magnitude has been increased by the factor $\alpha_{\text{lag}} = 10.15$, which allows us to meet the steady-state error specification. The phase lag introduced by the compensator has mostly dissipated by $\omega_w = 1.747$ rad/s, so the phase margin of the final design (solid curve) is 50°, about 5° less than that for the proportional controller (dashed curve).

The transfer function from the reference input to the output is given by

$$T(z) = \frac{G(z)G_c(z)}{1 + \overline{GH}(z)G_c(z)} \tag{9.19}$$

Figure 9.6 shows the step-reference response, which can be compared with that of the PI design done in Example 8.2 and illustrated in Figure 8.5. We see that this frequency-response design has resulted in a shorter rise time with about the same amount of overshoot for the reference response. There is a 15% undershoot after the initial peak, a transient that decays more slowly than in the PI response, and a steady-state error of 2%, however. These are typical tradeoffs that control-system designers must make, and there are no easy ways to resolve them.

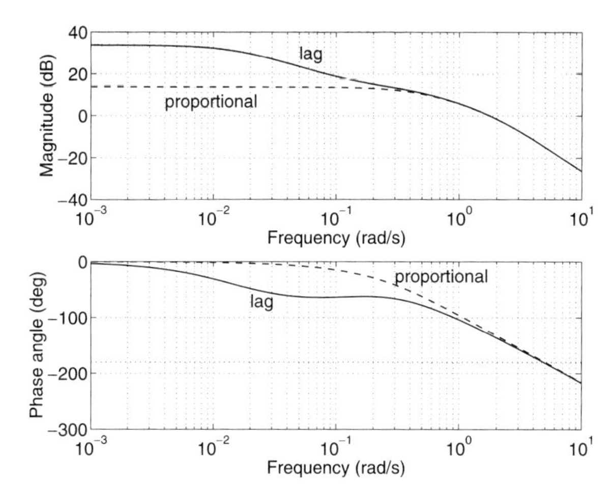

FIGURE 9.5 *Bode plots for initial (proportional) and final (lag) designs in Example 9.2*

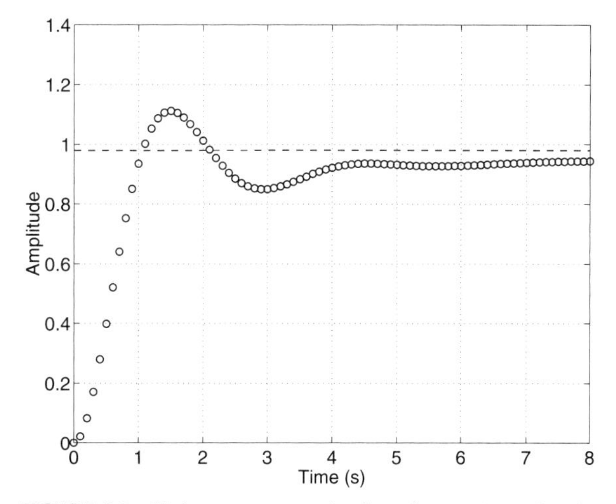

FIGURE 9.6 *Unit-step response in the reference input for the final lag design of Example 9.2*

□

LEAD CONTROLLER DESIGN

Like its counterpart, the proportional-derivative (PD) controller, a lead controller is generally used to reduce the rise time or, equivalently, to increase the bandwidth of the closed-loop frequency response for the feedback system shown in Figure 9.3. As we did with the lag controller, we assume that specifications are to be met for the steady-state error, which will establish the minimum required low-frequency gain, and for the phase margin, which should ensure adequate damping. Keep in mind that, although we will be satisfying a specification on the steady-state error caused by a reference-step input, only the gain of the lead is used to meet that requirement. The dynamic aspects of the lead, as established by its zero and pole (or zero and α) are used to provide the required phase margin, subject to the constraint that the lead gain be established based on steady-state error considerations.

The method described by Franklin, Powell, and Emami-Naeini (1994), begins by fixing the controller gain K_{lead} at the value that provides the required low-frequency gain. Presumably this gain results in a phase margin that is too low to meet the specification and yields a step response that is too lightly damped. The system could even be unstable at this point in the design process.

To rectify this situation, a lead compensator is added with the transfer function

$$G_c(w) = K_{\text{lead}}\alpha_{\text{lead}}\left(\frac{w - z_{\text{lead}}}{w - \alpha_{\text{lead}}z_{\text{lead}}}\right) \tag{9.20}$$

This transfer function's low-frequency gain is K_{lead}. The lead compensator, or controller, provides phase lead and increases the mid- and high-frequency magnitudes. Whereas the former is beneficial to the phase margin, the latter is not. Thus, the designer must select the parameter values (z_{lead} and α_{lead}) to balance the positive and negative effects.

After the gain K_{lead} has been added, we use MATLAB to compute the modified phase margin, expecting that it will not satisfy the specification. Then we decide on a phase-margin target value, realizing that we will not end up with anything close to that value because the magnitude-versus-frequency curve generally has a negative slope, and the gain-crossover frequency ω_{wc} is going to increase substantially because of the lead. The desired increase in the phase margin, denoted as $\Delta\phi_m$, is the difference between the user-specified target value and the actual phase margin with just K_{lead} included. It is related to the lead's pole-zero ratio according to

$$\alpha_{\text{lead}} = \frac{1 + \sin\Delta\phi_m}{1 - \sin\Delta\phi_m} \tag{9.21}$$

which can be used to solve for α_{lead}. Also, we know that the maximum phase lead will occur at the lead's center frequency, which is defined as

$$\omega_{w\text{ctr}} = z_{\text{lead}}\sqrt{\alpha_{\text{lead}}} \tag{9.22}$$

At this frequency the magnitude of the lead's frequency response is $\sqrt{\alpha_{\text{lead}}}$.

At this point we use the MATLAB `interp1` command with the open-loop frequency response data obtained with only K_{lead} as the controller to solve for the center frequency of the lead ω_{ctr} to satisfy the relationship

$$|K_{\text{lead}}\overline{GH}(j\omega_{w\text{ctr}})| = 1/\sqrt{\alpha_{\text{lead}}} \tag{9.23}$$

As a consequence, $\omega_{w\text{ctr}}$ becomes the gain-crossover frequency when the lead is installed, because the magnitude of the series combination of the lead and the plant (including any sensor) is unity (or 0 dB) at this frequency. Then we use (9.22) to solve for the lead zero.

When we compute the system's phase margin with the lead compensator in place, we generally find that we achieve only a fraction of the increase in phase margin that we had asked for. As described above, this partial success is because the gain-crossover frequency ω_{wc} is being moved to a higher frequency, and the phase of the plant's frequency response is decreasing. Thus in practice, we ask for a somewhat larger increase in phase margin than we really need, and we expect to make several attempts before arriving at a satisfactory choice. We illustrate these points in the following example.

☐ EXAMPLE 9.3 *Lead Controller Design*

Use the frequency-response approach described above to design a lead compensator for the plant and sensor whose transfer functions are given in Example 9.2—namely,

$$G_p(s) = \frac{4}{(2s+1)(0.5s+1)} \quad \text{and} \quad H(s) = \frac{1}{0.05s+1}$$

and a sampling period of $T_s = 0.1$ s. The closed-loop sampled-data system should have a steady-state error of 10% to a unit-step reference input and a phase margin of at least 50°.

Solution

The solution for this example is divided into the four steps described below, and the MATLAB commands that implement these steps are given in Script 9.3.

a. Because the plant and sensor have no poles at $s = 0$, they constitute a type-0 system. The steady-state error specification of $10\% = 1/10$ dictates a position constant, as defined by Equation (7.3), of $K_p = 10 - 1 = 9$. The DC gain of the plant and sensor is $\overline{GH}(w)|_{w=0} = 4$, so the lead compensator must supply a gain of $9/4 = 2.25$ at low frequencies, which requires that $K_{\text{lead}} = 2.25$.

b. Next we compute the phase margin with the proportional controller $G_c(w) = K_{\text{lead}}$ and find it to be 32.8°, which is below the specification of 50°. To compensate for the decreasing magnitude and phase of the plant in the gain-crossover region, we ask for a total phase margin of 65°—a phase increase of $\Delta\phi_m = 65 - 32.8 = 32.2°$. Then we use (9.21) to solve for $\alpha_{\text{lead}} = 3.28$.

c. Because we want the center frequency of the lead compensator to be the new gain-crossover frequency ω_{wc}, we look for the frequency for which the magnitude response of the open-loop system with the gain $K_{\text{lead}} = 2.25$ included is $1/\alpha_{\text{lead}} = 0.3047$. The **bodedb** function is used to generate the array **mag_db**, the magnitude in db of the frequency response, which is then converted to `mag_ratio`. By doing an interpolation with the vectors **ww** and `mag_ratio`, we find the gain-crossover frequency to be $\omega_{\text{ctr}} = 5.244$ rad/s. Based on these values for the center frequency and for α_{lead}, the zero and pole of the lead compensator are $z_{\text{lead}} = -\omega_{wc}/\sqrt{\alpha_{\text{lead}}} = -2.895$ and $p_{\text{lead}} = z_{\text{lead}}\alpha_{\text{lead}} = -9.5$. With these choices, we compute the actual phase margin as $51.1°$, which satisfies the phase-margin design objective. The overall controller is given by

$$G_c(w) = 7.384\left(\frac{w + 2.895}{w + 9.5}\right)$$

The Bode plots of the compensated open-loop frequency response in Figure 9.7 show how the lead has increased the magnitude at the higher frequencies and has increased the phase angle over the mid to high frequencies.

d. The digital controller is obtained by using the **c2d** command with the `tustin` option to apply the bilinear transform (9.6) to $G_c(w)$, resulting in

$$G_c(z) = 5.731\left(\frac{z - 0.7471}{z - 0.3559}\right)$$

The discretized plant $G_p(z)$ is obtained using the **c2d** command with the **zoh** option, and the closed-loop transfer function is computed via the "/" operator to get **Tz**.

We complete the design process by plotting the response of the closed-loop system to a unit-step reference input; this is shown in Figure 9.8. When this response is compared with the step response in Figure 9.6, which was obtained with a lag controller and has steady-state errors of only 2% to the step-reference input, we see that the lead controller has resulted in an appreciably faster response but five times the steady-state error. In general, we would use a lead to get a faster response and a lag to get lower steady-state errors. In the next section we illustrate the design of a lead-lag controller that combines these features and results in corresponding improvements in the reference step response.

MATLAB Script

```
% Script 9.3 -- lead controller design
%       use GHw, Ts, and Gp from Script 9.2(a)
%----- Part (a) determine lead gain -----
Kp_reqd = (1/0.10) - 1                % required gain to give 4% ess
plant_lfg = dcgain(GHw)
Klead = Kp_reqd/plant_lfg
%----- Part (b) determine lead alpha -----
[gm,pm,wwgm,wwpm] = margin(Klead*GHw)
pm_target = 65                        % phase-margin target in degrees
del_pm = pm_target - pm               % max phase angle in degrees
Alead = (1+sin(del_pm*pi/180))/(1-sin(del_pm*pi/180)) % Eq (9.21)
%----- Part (c) determine lead zero -----
ww = logspace(-1,2,100);              % set up frequency scale
[mag_db,ph] = bodedb(Klead*GHw,ww)    % compute frequency response
mag_ratio = 10.^(mag_db/20)           % convert to ratio
```

```
wwc = interp1(mag_ratio,ww,1/Alead)   % find wwc
Zlead = -wwc/sqrt(Alead)              % compute zero
Gc1 = tf(Alead*[1 -Zlead],[1 -Alead*Zlead]) % lead section
Gcw = Klead*Gc1                       % lead controller
GcGHw = Gcw*GHw                       % OL tf of lead, plant, & sensor
[kmc,pmc,wwkm,wwpm] = margin(GcGHw)   % find new phase margin
%----- Part (d) CL system for ref input with sensor in fdbk path -----
Gcz = c2d(Gcw,Ts,'tustin')            % convert to DTS controller
Gz = c2d(Gp,Ts,'zoh')
Tz = Gz*Gcz/(1+GHz*Gcz)               % closed-loop transfer function
step(Tz,3)                            % step response to 3 s
```

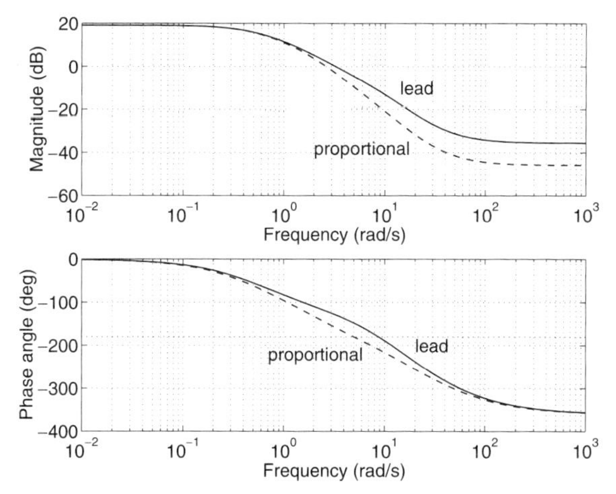

FIGURE 9.7 *Bode plots of initial (proportional) and final (lead) designs for Example 9.3*

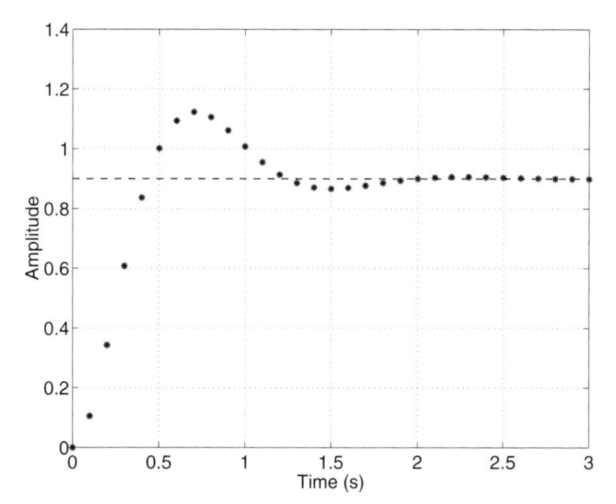

FIGURE 9.8 *Unit-step response for the final lead design of Example 9.3*

□

Typically, lead-lag designs are appropriate when we require a small steady-state error, a short rise time, and a small overshoot. It generally is not possible to meet such specifications with either a lead or a lag alone, but it often is possible if *both* features are included in the controller. Combining a lead and a lag section, we write the controller transfer function as

$$G_c(w) = K_{\text{ldlg}} \left(\frac{\alpha_{\text{lead}}(w - z_{\text{lead}})}{w - \alpha_{\text{lead}} z_{\text{lead}}} \right) \left(\frac{(w - z_{\text{lag}})/\alpha_{\text{lag}}}{w - z_{\text{lag}}/\alpha_{\text{lag}}} \right) \tag{9.24}$$

where both α_{lead} and α_{lag} are greater than unity.

To carry out the design, we must select the controller gain (K_{ldlg}), the zero in the w-plane and α for the lead section ($z_{\text{lead}}, \alpha_{\text{lead}}$), and the zero in the w-plane and α for the lag section ($z_{\text{lag}}, \alpha_{\text{lag}}$). From experience, we expect that $z_{\text{lead}} < z_{\text{lag}} < 0$.

A lead-lag controller simplifies the design process because there are more design parameters available than when using a lead or a lag alone. We use this freedom when selecting the gain-crossover frequency, which directly affects the response time and the closed-loop bandwidth.

Consider a control design requiring a certain steady-state error and phase margin ϕ_m. First we compute the gain of the controller to satisfy the system low-frequency gain requirement. Then we plot the frequency response of the open-loop system with the controller gain K_{ldlg}, from which we decide on a suitable gain-crossover frequency, ω_{wc}. This is followed by setting the center frequency of the lead section at ω_{wc} so that the phase of the compensated system is $-180° + \phi_m + 5°$, where the $5°$ term is introduced to compensate for the phase lag from the lag section. Finally, the lag section is designed to reduce the gain at ω_{wc} to be equal to 0 dB while keeping the DC gain of the lag section at unity. The design procedure is illustrated in the next example.

☐ EXAMPLE 9.4 *Lead-Lag Controller Design*

Find a lead-lag controller that results in a steady-state error of 1% and a phase margin of at least 50° for the plant and sensor used in Examples 9.2 and 9.3. These transfer functions are

$$G_p(s) = \frac{4}{(2s+1)(0.5s+1)} \quad \text{and} \quad H(s) = \frac{1}{0.05s+1}$$

and the sampling period is $T_s = 0.1$ s. The values of α for both the lead and the lag should not exceed 20. After you complete the design, determine the transfer function of the closed-loop system and find its bandwidth.

Solution

The solution for this example is divided into the four steps described below, and the MATLAB commands that implement these steps are given in Script 9.4.

a. Because $G_p(0) = 4$ and $H(0) = 1$, the controller gain that meets the steady-state error specification is $K_{\text{ldlg}} = 100/4 = 25$. The frequency-response plot of $K_{\text{ldlg}}\overline{GH}(w)$ is shown as the dashed line in Figure 9.9. In Example 9.3 the lead controller achieved a gain-crossover frequency of about 3.30 rad/s. Given the additional freedom in the lead-lag design, we increase the desired gain-crossover frequency to 5 rad/s, to achieve a faster time response.

b. Next we insert the lead section with unity DC gain and a center frequency of 5 rad/s. To determine α_{lead}, we compute the phase of the frequency response of $K_{\text{ldlg}}\overline{GH}(w)$ at 5 rad/s, which we find to be $-179.8°$. Because the desired phase margin is $50°$, the maximum phase lead needs to be $50° + 5° - 180° + 179.8° = 54.8°$. Applying (9.21), we find α_{lead} to be 9.93. From (9.22) the lead pole is at $w = -5/\sqrt{\alpha_{\text{lead}}} = -1.587$, and the lead zero is at $w = -5 \times \sqrt{\alpha_{\text{lead}}} = -15.76$.

c. Cascading the controller gain and lead controller with the plant model, we find the magnitude of the frequency response of the resulting system at 5 rad/s to be 11.63. Thus the lag section must reduce the gain at 5 rad/s by a factor of 11.63. From the discussion of the lag design in Example 9.2, we select the zero of the lag to be $z_{\text{lag}} = -0.5$, whose magnitude is one-tenth of 5 rad/s, and let $\alpha_{\text{lag}} = 11.63$, so that the lag pole is at $w = -0.5/11.63 = -0.04301$. After the gain, the lead section, and the lag section are combined, the transfer function of the continuous-time w-plane controller is

$$G_c(w) = 21.36 \left(\frac{w + 1.587}{w + 15.76}\right)\left(\frac{w + 0.5}{w + 0.04301}\right) \tag{9.25}$$

With the controller $G_c(w)$, we obtain a phase margin of $49.6°$, which is considered to be satisfactory for the design. The frequency response of the compensated system $\overline{GH}(w)G_c(w)$ is plotted as the solid line in Figure 9.9. Note the effect of the controller on the phase of the open-loop compensated system.

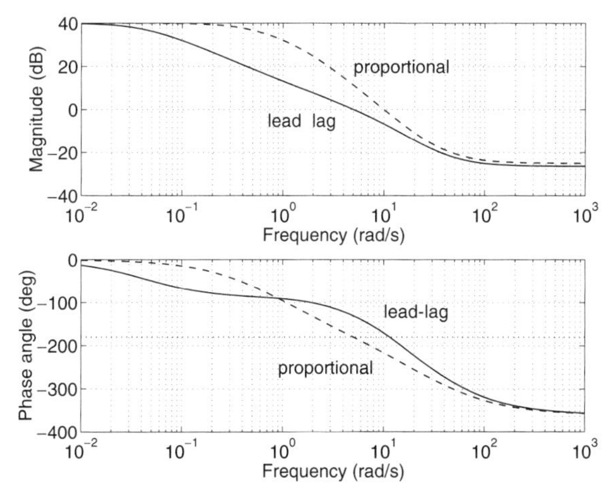

FIGURE 9.9 *Bode plots of initial proportional (dashed) and lead-lag (solid) designs for Example 9.4*

MATLAB Script

```
% Script 9.4  lead-lag controller design
%      use Gp, GHz, GHw, & Ts from Script 9.2(a)
%----- Part (a) set gain and make frequency response plot -------
Kp_reqd = 100                             % will give ess = 1/101
plant_lfg = dcgain(GHw)
Kldlg = Kp_reqd/plant_lfg
bode(Kldlg*GHw)                           % make open-loop freq resp plot
%----- Part (b) insert lead, determine its alpha ------
ww1 = 5                                   % lead center frequency
pm_target = 55                            % phase-margin target in degrees
[mag1,ph1] = bodedb(Kldlg*GHw,ww1)
del_pm = pm_target - (180 + ph1)          % max phase angle in degrees
Alead = (1+sin(del_pm*pi/180))/(1-sin(del_pm*pi/180)) % Eq (9.21)
Zlead = -ww1/sqrt(Alead)                  % lead zero
Gcldw = Kldlg*tf(Alead*[1  -Zlead],[1  -Alead*Zlead])  % lead
%------ Part (c) insert lag & determine its alpha -----------
[mag2,ph2] = bode(Gcldw*GHw,ww1)          % find gain at crossover
Alag = mag2                               % lag alpha
Zlag = -ww1/10                            % lag zero
Gclgw = tf([-1/Zlag  1],[-Alag/Zlag  1]) % unity DC gain lag
Gcldlgw = Gcldw*Gclgw  % combine lead & lag for final lead-lag controller
GcldlgGHw = Gcldlgw*GHw    % put controller in series with plant + sensor
[km,pm,wkm,wpm] = margin(GcldlgGHw)       % want pm = 50
%---- Part (d) build CL system for ref input with sensor in fdbk path
Gcldlgz = c2d(Gcldlgw,Ts,'tustin')        % convert Gc(w) to Gc(z)
Gcldlgz = ss(Gcldlgz)                     % convert to ss form
Gz = c2d(Gp,Ts,'zoh')                     % obtain G(z)
Tz = Gcldlgz*Gz/(1+Gcldlgz*GHz)           % closed-loop system
step(Tz,3)                                % step response to 3 s
%----  bandwidth of CL system using narrow freq interval
ww = logspace(0,1.5,100)';                % 1.5-decade freq interval
[mag_db_CL,ph_CL] = bodedb(Tz,ww);        % frequency response
lfg = dcgain(Tz)                          % low-freq gain
lfg_db = 20*log10(lfg)                    % low-freq gain in dB
bw_CL = bwcalc(mag_db_CL,ww,lfg_db)       % calc CL bandwidth in rad/s
```

When $G_c(w)$ is converted to the discrete-time domain, the transfer function of the discrete-time controller is

$$G_c(z) = 13.19 \left(\frac{z - 0.9512}{z - 0.9957} \right) \left(\frac{z - 0.8530}{z - 0.1187} \right) \tag{9.26}$$

d. The last block of commands in the script computes the bandwidth of the closed-loop system. To accomplish this we (i) build the model of the closed-loop system, (ii) select a one-and-a-half-decade–wide range of frequencies that includes the point at which the magnitude is 3 decibels below the low-frequency value and evaluate the closed-loop frequency response, (iii) use the `dcgain` command to find the low-frequency gain (−0.0864 dB), and (iv) use the RPI function `bwcalc` to compute the bandwidth, which we find to be 10.51 rad/s. One might be tempted to use the interpolation command `interp1` instead. The magnitude function `mag_CL_db` is not monotonic, however, which prevents the use of `interp1` unless the range of frequencies is suitably restricted.

The reference step response for the final design is shown in Figure 9.10. Comparing the reference step response with that of Figure 8.8, which was obtained with a PID controller, we see that in this case there is more overshoot and a longer settling time. This design has a steady-state error of $1/101 = 0.99\%$, whereas the PID design from Chapter 8 has zero steady-state error. With the lead-lag design, no effort was made to optimize performance measures such as the overshoot and settling time by making different choices for the phase margin.

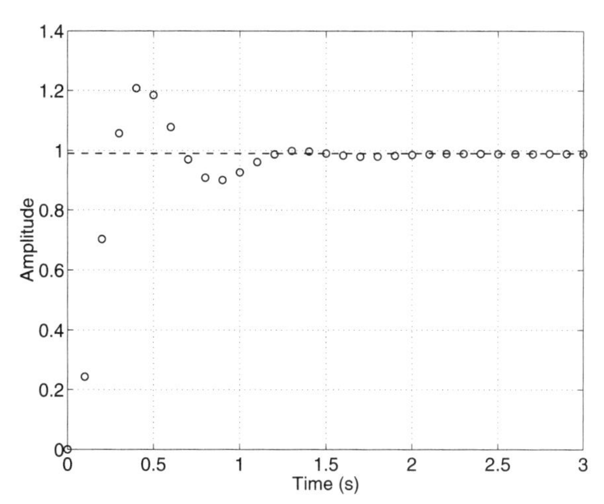

FIGURE 9.10 *Unit-step response for the lead-lag design of Example 9.4*

WHAT IF?

a. Repeat the design for Example 9.4 using an allowance of 6° (instead of 5°) in defining the pole-zero ratio α_{lead}. You should find that the controlled system has a phase margin slightly greater than 50°.

b. Repeat the design for Example 9.4 with the desired phase margin increased to 60°. This design should result in smaller overshoot in the step response. ■ □

EXPLORATORY PROBLEMS

EP9.1 Lag compensator for type-0 plant. Consider the fourth-order continuous-time plant having the transfer functions

$$G_p(s) = \frac{160(s+10)}{(s+2)(s+4)(s+5)(s+20)} \quad \text{and} \quad H(s) = 1$$

for the sampled-data control system of Figure 9.1, with a sampling period $T_s = 0.02$ s. Using the frequency-response design approach illustrated in Example 9.2, design a lag controller that results in a phase margin of at least 60° and a steady-state error of no more than 2%.

COMPREHENSIVE PROBLEMS

CP9.1 Ball and beam system. Use the frequency-response approach to design a lead controller, $K_b(z)$, for the ball and beam system in Figure A.3 of Appendix A, with $K_m(z)$ set to be a proportional controller with a gain of 5. First close the $K_m(z)$ (wheel-angle) loop to obtain the system $G_b(z)$, and then convert $G_b(z)$ to $G_b(w)$ for designing the controller $K_b(w)$. Before performing the design, find the gain margin of $G_b(w)$ under unity feedback (i.e., with $K_b(w) = 1$) and simulate the step response of the unity feedback system. Note that the ball control loop has a small phase margin (about 15°) and the time response is oscillatory, with the first peak at about 7 s.

To improve the response speed and damping, a lead controller for $K_b(w)$ is needed. For this position loop, it is desired to have a gain margin of 10 dB, a phase margin of 45°, and a gain-crossover frequency of 1 rad/s. Convert the controller $K_b(w)$ back to a discrete-time controller, $K_b(z)$. To evaluate $K_b(z)$, apply a step input of magnitude 0.1 m to the feedback system and determine the rise time, overshoot, and settling time of the ball position. The model for the ball and beam system discretized at 50 Hz can be obtained from the file `dbbeam.m`.

CP9.2 Electric power system. Use the frequency-response method discussed in this chapter to design a lag controller $K_V(z)$ for the voltage regulation of the power system model shown in Figure A.8 of Appendix A. As in Comprehensive Problem CP8.3, use the model file `dpower2.m`, with a sampling frequency of 100 Hz. Apply the bilinear transform to the discretized model to obtain an equivalent model in the w-domain.

a. Design a lag controller $K_V(w)$ such that the steady-state error to a unit-step input V_{ref} of magnitude 0.05 is no more than 0.001 and the phase margin is at least 90°. The phase margin is set quite high because if it were set lower, the gain-crossover frequency would be close to that of the electromechanical mode and thus would reduce its damping. Convert $K_V(w)$ back to $K_V(z)$ and obtain the closed-loop transfer function of the discrete-time system. Plot the response of the discrete-time closed-loop system subject to an input voltage step of magnitude 0.05. From the step response, verify that the steady-state error specification is satisfied, and find the rise time, maximum overshoot, and 2% settling time.

b. Repeat the design for phase margins of 85° and 95°. For the 85° design, note that the time response is faster but the oscillation is less damped, whereas the reverse is true for the 95° design.

CP9.3 Hydroturbine system. Consider the design of a lag controller for the hydroturbine model in Figure A.10 of Appendix A. Discretize the model at 10 Hz and use the bilinear transform to convert the model to $G(w)$ in the w-domain. Design $G_c(w)$ so that the steady-state error is less than 5% for a unit-step power command P_{des} and the phase margin is at least 45°. Obtain $G_c(z)$ from $G_c(w)$ and build the closed-loop discrete-time system. Plot the unit-step response of the closed-loop system and verify the design results. Note that if the design steps for the lag controller are carried out directly on $G(w)$, the result will not be acceptable because of a peak in the amplitude of the open-loop frequency response at about 1.2 rad/s. If we apply a pre-filter $G_f(w) = 0.05/(w + 0.05)$ to smooth out the peak in $G(w)$, the lag controller design algorithm can be readily applied to $G(w)G_f(w)$. The resulting overall controller will be $G_c(w)G_f(w)$.

SUMMARY

In this chapter we have shown how MATLAB can be used to carry out the calculations required to design lead-lag controllers using frequency-response methods. The approach used here requires the bilinear transformation of the discretized model to form an equivalent continuous-time system. Continuous-time-domain design techniques for lead, lag, and lead-lag controllers are then applied to obtain controllers to satisfy specifications such as gain and phase margins. Examples were given to illustrate the design procedures.

10 *State-Space Design*

PREVIEW

In contrast to an input-output transfer-function model, a state-space model contains information on the internal states. Time-domain control design methods using state-space models are largely based on utilizing the internal states as part of the design scheme. In pole placement, a primary design technique, the closed-loop system poles are placed at specific locations in the z-plane. If all the states are available for feedback control, the design reduces to the computation of a static feedback-gain matrix. If some of the states are not measurable, a state estimator or observer can be constructed. The separation principle allows for independent computation of the full-state feedback-gain matrix and the state-estimator gain matrix. In this chapter we illustrate these design methods. To complete the overall picture of the design process, we also discuss the determination of system controllability and observability.

CONTROLLABILITY

Consider the state-space model

$$\mathbf{x}(k+1) = \mathbf{A}\mathbf{x}(k) + \mathbf{B}u(k), \quad \mathbf{y}(k) = \mathbf{C}\mathbf{x}(k) \tag{10.1}$$

where $\mathbf{x}(k)$ is the n-vector of state variables, $u(k)$ is the control variable, and $\mathbf{y}(k)$ is the p-vector of output variables. In this chapter we consider only single-input systems but allow for multiple outputs. In addition, the $\mathbf{D}$ matrix that relates the input $u(k)$ to the output $\mathbf{y}(k)$ is assumed to be zero and is not shown in (10.1).[1] MATLAB commands that operate on state matrices will require it as an argument, however.

The system (10.1) is said to be controllable if there exists a constant gain matrix $\mathbf{F}$ such that the feedback control law

$$u(k) = -\mathbf{F}\mathbf{x}(k) \tag{10.2}$$

places the poles of the closed-loop system—that is, the eigenvalues of $\mathbf{A} - \mathbf{BF}$—at any arbitrary locations. A test for controllability is that $\text{rank}(\mathcal{C}) = n$, where $\mathcal{C}$ is the controllability matrix defined as

$$\mathcal{C} = \begin{bmatrix} \mathbf{B} & \mathbf{AB} & \cdots & \mathbf{A}^{n-1}\mathbf{B} \end{bmatrix}$$

If $\text{rank}(\mathcal{C}) = r < n$, only r eigenvalues of $\mathbf{A} - \mathbf{BF}$ can be arbitrarily assigned.

The following example illustrates the use of the Control System Toolbox `ctrb` function to generate $\mathcal{C}$ and to determine how many eigenvalues are controllable.

☐ EXAMPLE 10.1 *Controllability Matrix*

Find the controllability matrix $\mathcal{C}$ for a system in SS form described by Equation (10.1) where

$$\mathbf{A} = \begin{bmatrix} 0.81 & -0.23 & -0.045 \\ 0.09 & 0.98 & -0.0023 \\ 0.005 & 0.10 & 1 \end{bmatrix} \quad \text{and} \quad \mathbf{B} = \begin{bmatrix} 0.09 \\ 0.0047 \\ 0.00016 \end{bmatrix}$$

and determine whether the system is controllable.

Solution Using the MATLAB commands in Script 10.1, we obtain `Co` as

$$\mathcal{C} = \begin{bmatrix} 0.0900 & 0.0718 & 0.0552 \\ 0.0047 & 0.0127 & 0.0189 \\ 0.0002 & 0.0011 & 0.0027 \end{bmatrix}$$

Because `c_eig`, the rank of `Co`, is 3, the system is controllable.

[1]The $\mathbf{D}$ matrix does not affect the determination of system controllability and observability, and the full-state feedback-gain and observer-gain design. It needs to be accounted for in the observer implementation, however.

MATLAB Script

```
% Script 10.1:  determine system controllability
A = [0.81   -0.23   -0.045
     0.09    0.98   -0.0023
     0.005   0.10    1      ]         % state matrix
B = [ 0.09 0.0047 0.00016]'           % input matrix
Co = ctrb(A,B)                        % controllability matrix
c_eig = rank(Co)                      % number of controllable eigenvalues
```

□

REINFORCEMENT PROBLEMS

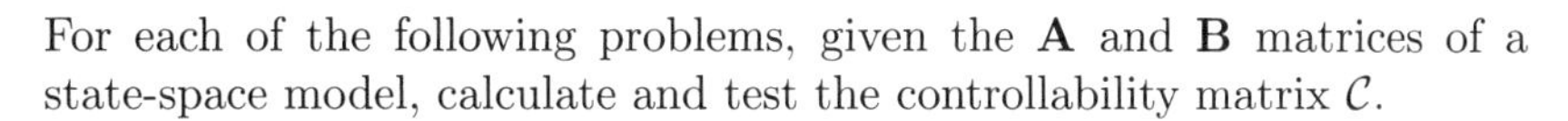

For each of the following problems, given the **A** and **B** matrices of a state-space model, calculate and test the controllability matrix $\mathcal{C}$.

P10.1 Third-order system.

$$\mathbf{A} = \begin{bmatrix} 0.8238 & 0.0400 & 0.0810 \\ 0.0459 & 0.7799 & 0.1161 \\ 0.0781 & 0.0020 & 0.7448 \end{bmatrix} \quad \text{and} \quad \mathbf{B} = \begin{bmatrix} 0.0503 \\ 0.0296 \\ 0.0887 \end{bmatrix}$$

P10.2 Fourth-order system in modal form.

$$\mathbf{A} = \begin{bmatrix} 0.9048 & 0 & 0 & 0 \\ 0 & 0.8187 & 0 & 0 \\ 0 & 0 & 0.7788 & 0 \\ 0 & 0 & 0 & 1 \end{bmatrix} \quad \text{and} \quad \mathbf{B} = \begin{bmatrix} 0.0476 \\ 0 \\ 0.0885 \\ 0.5 \end{bmatrix}$$

Given a system in modal form, show that any distinct eigenvalue in **A** corresponding to a zero entry in **B** is not controllable.

P10.3 Controller form.

$$\mathbf{A} = \begin{bmatrix} 2.2 & -1.69 & 0.528 & -0.054 \\ 1 & 0 & 0 & 0 \\ 0 & 1 & 0 & 0 \\ 0 & 0 & 1 & 0 \end{bmatrix} \quad \text{and} \quad \mathbf{B} = \begin{bmatrix} 1 \\ 0 \\ 0 \\ 0 \end{bmatrix}$$

This system is in controller canonical form. Verify that its controllability matrix $\mathcal{C}$ is upper triangular with the 1's on the diagonal. Thus $\mathcal{C}$ has full rank, and the system is always controllable, regardless of the values of the entries in the first row of **A**.

POLE PLACEMENT

Because system transient behaviors caused by initial conditions and inputs depend directly on the system poles, the objective of a pole-placement design is to apply feedback such that the system transients decay in an acceptable time interval. When a system is controllable and all its states are available for feedback, a full-state feedback control (10.2) can be applied to place the poles of the closed-loop system at arbitrary locations in the z-plane.

The algorithm commonly cited in control textbooks for pole-placement design of single-input systems is the Ackermann formula, which is available in the Control System Toolbox as the `acker` function. The algorithm involves the inversion of the controllability matrix, however, which can be ill-conditioned for high-order systems. The Control System Toolbox `place` function utilizes a more robust pole-placement algorithm that is also applicable to multi-input systems and is recommended for general use (even for single-input systems). The use of `place` is illustrated in the next example.

☐ EXAMPLE 10.2 *Pole Placement by Full-State Feedback*

Find the control-gain vector $\mathbf{F}$ such that when the system given in Example 10.1 is controlled by Equation (10.2), the closed-loop poles are at $z = 0.75, 0.80$, and 0.90. Verify the control gain by finding the eigenvalues of $\mathbf{A} - \mathbf{BF}$ and comparing them with the desired closed-loop poles.

Solution

In using the `place` command, we specify the matrices $\mathbf{A}$ and $\mathbf{B}$ and the desired closed-loop poles as the column vector $\mathbf{p} = \begin{bmatrix} 0.75 & 0.80 & 0.90 \end{bmatrix}^T$. (Note that $\mathbf{p}$ can also be a row vector.) The commands in Script 10.2 find the required control gain vector and verify the result. The computation shows that

$$\mathbf{F} = \begin{bmatrix} 3.4040 & 6.9869 & 5.0242 \end{bmatrix}$$

Note that the closed-loop system matrix

$$\mathbf{A}_{\text{cl}} = \mathbf{A} - \mathbf{BF} = \begin{bmatrix} 0.5036 & -0.8588 & -0.4972 \\ 0.0740 & 0.9472 & -0.0259 \\ 0.0045 & 0.0989 & 0.9992 \end{bmatrix}$$

has no particular structure. The computation of the eigenvalues of $\mathbf{A}_{\text{cl}}$ indicates that $\mathbf{F}$ has been computed accurately. For high-order systems, it is a good idea to check the eigenvalues of the closed-loop system to verify the accuracy of the closed-loop poles. We can evaluate the accuracy of the calculations by computing the eigenvalues of $\mathbf{A}_{\text{cl}}$ and subtracting them from $\mathbf{p}$. Doing this, we find that the greatest difference in any of the three pairs of eigenvalues is 0.4×10^{-14}.

MATLAB Script

```
% Script 10.2:  pole-placement design
A = [0.81  -0.23   -0.045
     0.09   0.98   -0.0023
     0.005  0.10    1     ]          % state matrix
B = [0.09 0.0047 0.00016]'           % input matrix
p = [0.75  0.80  0.90]'              % desired pole locations
F = place(A,B,p)                     % control gain
A_cl = A-B*F                         % closed-loop system matrix
diff = sort(p)-sort(eig(A_cl))       % check closed-loop eigenvalues
```

□

Although controllability allows unlimited ability to shift the poles of a closed-loop system, the reality is that an actuator's control action is always bounded. An extremely large controller gain that, because of saturation, cannot be fully exercised by an actuator may even result in an unstable closed-loop system. Therefore, a practical pole-placement design must ensure that if a sufficiently large control gain is used, the saturation of the actuator does not cause undesirable system behavior. This design consideration encourages the shifting of the closed-loop poles to desirable locations within reasonable proximity of the open-loop poles. To measure the magnitude of a gain vector, $\mathbf{F}$, we use the 2-norm, which is defined as

$$\|\mathbf{F}\|_2 = \left(\sum_{i=1}^{n} f_i^2 \right)^{1/2}$$

where f_i, $i = 1, ..., n$, are the entries of $\mathbf{F}$. The norm of $\mathbf{F}$ can be computed using the MATLAB `norm(F)` command. The following example illustrates a low-gain design.

□ EXAMPLE 10.3 *Pole Placement (Low Gain)*

The open-loop system given in Example 10.1 has poles at $z = 0.9784$ and $z = 0.9124\epsilon^{\pm j0.1204}$. Determine the control gain $\mathbf{F}$ to provide a closed-loop system having poles at $z = 0.95$ and $z = 0.9055\epsilon^{\pm j0.1107} = 0.90 \pm j0.10$. These poles result in a stable closed-loop system that responds faster than it would in the absence of feedback, but the controller gains should be lower than those in Example 10.2 because the desired poles are closer to the open-loop poles. Compare the norm of $\mathbf{F}$ to that obtained in the previous example.

Solution

We use the commands in Script 10.2 with

$$\mathbf{p} = \left[\begin{array}{ccc} 0.95 & 0.90 + j0.10 & 0.90 - j0.10 \end{array} \right]^T$$

to find the desired control gain

$$\mathbf{F} = \left[\begin{array}{ccc} 0.4192 & 0.4635 & 0.6065 \end{array} \right]$$

The `norm(F)` command yields a value of 0.8709. In Example 10.2, the poles were shifted closer to the origin in the z-plane, resulting in $\mathbf{F}$ having a significantly higher norm, 9.255. □

REINFORCEMENT PROBLEMS

For each of the following problems, find the feedback-gain matrix $\mathbf{F}$ to place the poles of the closed-loop system at the specified locations. Verify that the closed-loop pole locations are those you requested.

P10.4 Third-order system. Place the poles of the system in Problem 10.1 at $z = 0.5$, 0.65, and 0.8.

P10.5 Fourth-order modal system. Place the poles of the system

$$\mathbf{A} = \begin{bmatrix} 0.9048 & 0 & 0 & 0 \\ 0 & 0.8187 & 0 & 0 \\ 0 & 0 & 0.7788 & 0 \\ 0 & 0 & 0 & 1 \end{bmatrix} \quad \text{and} \quad \mathbf{B} = \begin{bmatrix} 0.0476 \\ 0.0476 \\ 0.0885 \\ 0.5 \end{bmatrix}$$

at $z = 0.7$, 0.75, 0.85, and 0.95.

P10.6 Lightly damped modes. Place the poles of the system

$$\mathbf{A} = \begin{bmatrix} 0.9641 & 0.2462 & 0.0303 \\ -0.2462 & 0.9641 & 0.1934 \\ 0 & 0 & 0.6065 \end{bmatrix} \quad \text{and} \quad \mathbf{B} = \begin{bmatrix} 0.0056 \\ 0.0528 \\ 0.3935 \end{bmatrix}$$

at (i) $z = 0.9453\epsilon^{\pm j0.250}$ and 0.5762, and (ii) $z = 0.1$, 0.2, and 0.3. Find and compare the norms of $\mathbf{F}$.

OBSERVABILITY

The system described by (10.1) is observable if there exists a constant estimator gain matrix $\mathbf{L}$ such that the eigenvalues of $\mathbf{A} - \mathbf{LC}$ can be assigned to any arbitrary locations. A test for observability is that $\text{rank}(\mathcal{O}) = n$, where $\mathcal{O}$ is the observability matrix defined as

$$\mathcal{O} = \begin{bmatrix} \mathbf{C} \\ \mathbf{CA} \\ \vdots \\ \mathbf{CA}^{n-1} \end{bmatrix}$$

and n is the system order. If $\text{rank}(\mathcal{O}) = r < n$, only r eigenvalues of $\mathbf{A} - \mathbf{LC}$ can be arbitrarily assigned.

The following example illustrates the use of the Control System Toolbox `obsv` function to generate $\mathcal{O}$ and to determine how many eigenvalues of $\mathbf{A} - \mathbf{LC}$ can be assigned.

□ EXAMPLE 10.4 *Observability Matrix*

Find the observability matrix $\mathcal{O}$ for system described by Equation (10.1) where

$$\mathbf{A} = \begin{bmatrix} 0.81 & -0.23 & -0.045 \\ 0.09 & 0.98 & -0.0023 \\ 0.005 & 0.10 & 1 \end{bmatrix} \quad \text{and} \quad \mathbf{C} = \begin{bmatrix} 1 & 3.5 & 3 \end{bmatrix}$$

and determine whether the system is observable.

Solution

Using the commands in Script 10.4, we obtain the observability matrix as

$$\mathcal{O} = \begin{bmatrix} 1.000 & 3.500 & 3.000 \\ 1.140 & 3.500 & 2.947 \\ 1.253 & 3.463 & 2.888 \end{bmatrix}$$

The rank of $\mathcal{O}$ is 3, which confirms that the system is observable.

MATLAB Script

```
% Script 10.4:  determine system observability
A = [0.81    -0.23    -0.045
     0.09     0.98    -0.0023
     0.005    0.10     1      ]          % state matrix
C = [1 3.5 3]                           % output matrix
Ob = obsv(A,C)                          % observability matrix
o_eig = rank(Ob)                        % number of observable eigenvalues
```

□

REINFORCEMENT PROBLEMS

For each of the following problems, given the **A** and **C** matrices of a state-space model, calculate and test the observability matrix $\mathcal{O}$.

P10.7 Third-order system.

$$\mathbf{A} = \begin{bmatrix} 0.8238 & 0.0400 & 0.0810 \\ 0.0459 & 0.7799 & 0.1161 \\ 0.0781 & 0.0020 & 0.7448 \end{bmatrix} \quad \text{and} \quad \mathbf{C} = \begin{bmatrix} 0 & 1 & 0 \end{bmatrix}$$

P10.8 Fourth-order modal system.

$$\mathbf{A} = \begin{bmatrix} 0.9048 & 0 & 0 & 0 \\ 0 & 0.8187 & 0 & 0 \\ 0 & 0 & 0.7788 & 0 \\ 0 & 0 & 0 & 1 \end{bmatrix} \quad \text{and} \quad \mathbf{C} = \begin{bmatrix} 1 & 2 & 0 & 1 \end{bmatrix}$$

Given a system in modal form, show that any distinct eigenvalue in **A** corresponding to a zero entry in **C** is not observable.

P10.9 Observer form.

$$\mathbf{A} = \begin{bmatrix} 2.2 & 1 & 0 & 0 \\ -1.69 & 0 & 1 & 0 \\ 0.528 & 0 & 0 & 1 \\ -0.054 & 0 & 0 & 0 \end{bmatrix} \quad \text{and} \quad \mathbf{C} = \begin{bmatrix} 1 & 0 & 0 & 0 \end{bmatrix}$$

This system is in observer canonical form. Verify that its observability matrix $\mathcal{O}$ is lower triangular with the 1's on the diagonal. Thus $\mathcal{O}$ has full rank, and the system is always observable, regardless of the values of the entries in the first column of $\mathbf{A}$.

OBSERVER DESIGN

If a system described by (10.1) is completely observable, a *state estimator*—also called an *observer*—that uses the input $u(k)$ and the output $\mathbf{y}(k)$ to estimate the state variable $\mathbf{x}(k)$ can be determined. Provided that $\mathbf{D} = \mathbf{0}$, the equations of the observer are

$$\hat{\mathbf{x}}(k+1) = \mathbf{A}\hat{\mathbf{x}}(k) + \mathbf{B}\mathbf{u}(k) + \mathbf{L}[\mathbf{y}(k) - \hat{\mathbf{y}}(k)] \qquad (10.3)$$

$$\hat{\mathbf{y}}(k) = \mathbf{C}\hat{\mathbf{x}}(k) \qquad (10.4)$$

where $\hat{\mathbf{x}}(k)$ is the observer state that provides an estimate of $\mathbf{x}(k)$. Equations (10.3) and (10.4) can be combined and rewritten as

$$\hat{\mathbf{x}}(k+1) = (\mathbf{A} - \mathbf{LC})\hat{\mathbf{x}}(k) + \mathbf{B}u(k) + \mathbf{L}\mathbf{y}(k)$$

If all the eigenvalues of $\mathbf{A} - \mathbf{LC}$ are inside the unit circle in the z-plane, after the initial observer transient has decayed, $\hat{\mathbf{x}}(k)$ will agree with $\mathbf{x}(k)$. This is known as *asymptotic tracking*. Transposing $\mathbf{A} - \mathbf{LC}$, we note that the resulting matrix $\mathbf{A}^T - \mathbf{C}^T\mathbf{L}^T$ is like the full-state feedback matrix $\mathbf{A} - \mathbf{BF}$ except that $\mathbf{A}^T$ and $\mathbf{C}^T$ replace $\mathbf{A}$ and $\mathbf{B}$, respectively, and the design gain is $\mathbf{L}^T$ instead of $\mathbf{F}$. Thus the `place` command can be used to design the observer gain $\mathbf{L}$.

In observer design, it is logical to require that the state estimation error $\mathbf{x}(k) - \hat{\mathbf{x}}(k)$ decay faster than the system transients. This is achieved by using a gain $\mathbf{L}$ that is sufficiently large to place the observer poles closer to the origin in the z-plane than the system poles. $\mathbf{L}$ should not be excessively large, however, because that would result in an overly wide bandwidth for the observer, which is undesirable because of the presence of noise or unmodeled high-frequency dynamics. In the following example, we use the `place` command to design an observer.

☐ EXAMPLE 10.5 *Observer Design*

To provide for good tracking of the states in Examples 10.1 and 10.4, design an observer that has its poles at $z = 0.50$, 0.60, and 0.75. Find the required observer gain $\mathbf{L}$ and verify the solution. Then simulate the observer when the input is zero and the initial conditions are $\mathbf{x}(0) = \begin{bmatrix} 1 & -0.75 & 0.5 \end{bmatrix}^T$ and $\hat{\mathbf{x}}(0) = \mathbf{0}$. Draw $\mathbf{x}(k)$ and $\hat{\mathbf{x}}(k)$ on one plot and $\mathbf{y}(k)$ and $\hat{\mathbf{y}}(k)$ on a separate plot. Assume a sampling period of $T_s = 0.1$ s to perform the simulation.

Solution

We begin by defining the observer poles as the column vector p. After using the `place` command to compute the required gain matrix, we transpose the resulting gain matrix to obtain $\mathbf{L}$. To find $\hat{\mathbf{x}}(k)$ and $\hat{\mathbf{y}}(k)$, we first need to simulate the system with the initial condition $\mathbf{x}(0)$ using `lsim` and then take the resulting output $\mathbf{y}(k)$ as the input to the observer and simulate the observer using `lsim` again. The commands in Script 10.5 perform these functions. Note that we do not need the $\mathbf{B}$ and $\mathbf{D}$ matrices of the state-space model for this problem, because the input is zero. These two matrices are required to build the LTI system object, however.

MATLAB Script

```
% Script 10.5:  observer design
Ts = 0.1
A = [0.81   -0.23   -0.045
     0.09    0.98   -0.0023
     0.005   0.10    1     ]                 % state matrix
B = [0.09 0.0047 0.00016]'                   % input matrix
C = [1 3.5 3]                                % output matrix
D = 0                                        % throughput matrix
p = [0.50 0.60 0.75]'                        % desired observer pole locations
L = place(A',C',p)'                          % estimator gain
A_ob = A-L*C                                 % observer system matrix
eig(A_ob)                                    % check eigenvalues
x0 = [1; -0.75; 0.5]                         % initial conditions
dtime = [0:Ts:6]';                           % time array
u = zeros(1,length(dtime));                  % zero input
G = ss(A,B,C,D,Ts)                           % build system as LTI object
[y,dtime,x] = lsim(G,u,dtime,x0);            % simulate system to get y(k)
G_ob = ss(A_ob,L,C,D,Ts)                     % build observer as LTI object
[y_hat,dtime,x_hat] = lsim(G_ob,y,dtime); % simulate observer with zero ICs
% plot system and observer states
plot(dtime,x,'o',dtime,x_hat(:,1),'+',dtime,x_hat(:,2),'x',...
     dtime,x_hat(:,3),'^'), grid             % Figure 10.1(a)
legend('x_1','x_2','x_3','x_1hat','x_2hat','x_3hat')   % add legend
plot(dtime,y,'o',dtime,y_hat,'*')            % Figure 10.1(b)
legend('y','yhat')                           % add legend
```

The computation shows that

$$\mathbf{L} = \begin{bmatrix} 1.7737 & -0.9762 & 0.8610 \end{bmatrix}^T$$

and

$$\mathbf{A} - \mathbf{LC} = \begin{bmatrix} -0.9637 & -6.4380 & -5.3661 \\ 1.0662 & 4.3966 & 2.9262 \\ -0.8560 & -2.9134 & -1.5829 \end{bmatrix}$$

The eigenvalues of $\mathbf{A} - \mathbf{LC}$ are found to be at $z = 0.50, 0.60$, and 0.75. Figure 10.1(a) shows the time response of $\mathbf{x}(k)$ and $\hat{\mathbf{x}}(k)$. Note that all the observer states in $\hat{\mathbf{x}}(k)$ initially move in directions opposite to the system states in $\mathbf{x}(k)$. After the initial observer transients have decayed in about 2 s, however, $\hat{\mathbf{x}}(k)$ tracks $\mathbf{x}(k)$. As shown in Figure 10.1(b), the output variable $\hat{\mathbf{y}}(k)$ initially moves in the direction of $\mathbf{y}(k)$ and the transients decay in less than 1 s.

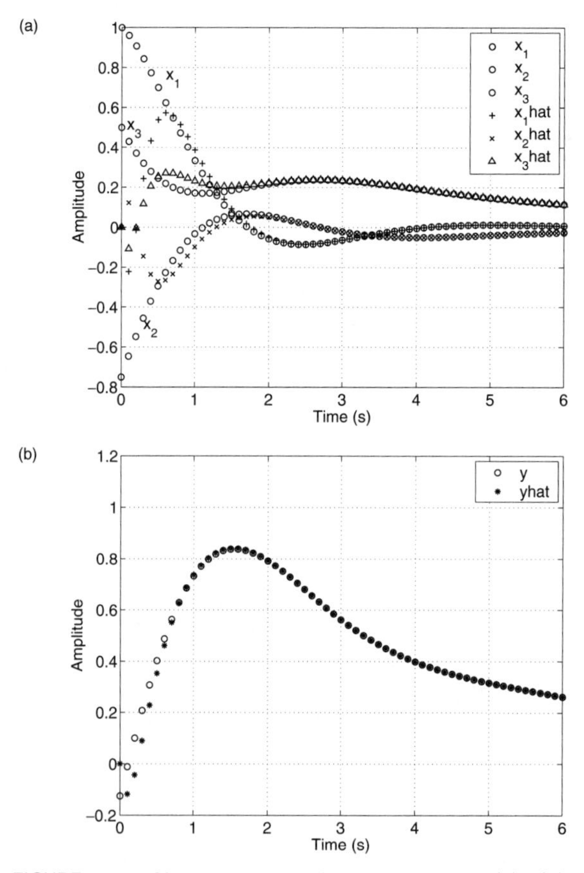

FIGURE 10.1 *Observer response for Example 10.5: (a)* $\mathbf{x}(k)$ *and* $\hat{\mathbf{x}}(k)$*; (b)* $\mathbf{y}(k)$ *and* $\hat{\mathbf{y}}(k)$

WHAT IF? Suppose that in Example 10.5 we desire the observer poles to be at $z = 0.3$, 0.4, and 0.65. Find the observer response and the norm of $\mathbf{L}$, and compare them to those obtained in Example 10.5. You will find that the new poles result in an $\mathbf{L}$ with a larger norm and larger initial transients in $\hat{\mathbf{x}}(k)$, but faster tracking of the system state $\mathbf{x}(k)$. ■

□

REINFORCEMENT PROBLEMS

In the following problems, given the $\mathbf{A}$ and $\mathbf{C}$ matrices of a state-space model, design an observer by using the gain $\mathbf{L}$ to place the poles of $\mathbf{A}-\mathbf{LC}$ at the specified locations. Simulate the observer given a zero input and the initial condition $\mathbf{x}(0)$. Assume $\hat{\mathbf{x}}(0) = \mathbf{0}$, and let $\mathbf{B}$ and $\mathbf{D}$ be zero. Make a single plot of $\mathbf{x}(k)$ and $\hat{\mathbf{x}}(k)$ and a second single plot of $\mathbf{y}(k)$ and $\hat{\mathbf{y}}(k)$. Assume a sampling time of $T_s = 0.1$ s to perform the simulation.

P10.10 Third-order system.

$$\mathbf{A} = \begin{bmatrix} 0.8238 & 0.0400 & 0.0810 \\ 0.0459 & 0.7799 & 0.1161 \\ 0.0781 & 0.0020 & 0.7448 \end{bmatrix} \quad \text{and} \quad \mathbf{C} = \begin{bmatrix} 0 & 1 & 0 \end{bmatrix}$$

Place the eigenvalues of $\mathbf{A} - \mathbf{LC}$ at $z = 0.3$, 0.4, and 0.6. The initial condition is $\mathbf{x}(0) = \begin{bmatrix} 1 & 0.5 & -0.25 \end{bmatrix}^T$.

P10.11 Fourth-order modal system.

$$\mathbf{A} = \begin{bmatrix} 0.9048 & 0 & 0 & 0 \\ 0 & 0.8187 & 0 & 0 \\ 0 & 0 & 0.7788 & 0 \\ 0 & 0 & 0 & 1 \end{bmatrix} \quad \text{and} \quad \mathbf{C} = \begin{bmatrix} 1 & 2 & -1 & 1 \end{bmatrix}$$

Place the eigenvalues of $\mathbf{A} - \mathbf{LC}$ at $z = 0.5$, 0.65, 0.78, and 0.92. The initial condition is $\mathbf{x}(0) = \begin{bmatrix} 1 & 2 & 0 & 1 \end{bmatrix}^T$.

P10.12 Lightly damped modes.

$$\mathbf{A} = \begin{bmatrix} 0.9641 & 0.2462 & 0.0303 \\ -0.2462 & 0.9641 & 0.1934 \\ 0 & 0 & 0.6065 \end{bmatrix} \quad \text{and} \quad \mathbf{C} = \begin{bmatrix} 1 & 0 & 0 \end{bmatrix}$$

Place the eigenvalues of $\mathbf{A} - \mathbf{LC}$ at $z = 0.3$, 0.6, and 0.7. The initial condition is $\mathbf{x}(0) = \begin{bmatrix} 1 & 0 & -2 \end{bmatrix}^T$.

The ability to assign the closed-loop poles to any location in the z-plane for a controllable system offers a strong motivation for using the full-state feedback gain as the controller. In realistic systems, however, not all the states are measured, for physical or economic reasons. On the other hand, the observer allows the reconstruction of the states from the outputs of the system and hence can be used to implement the full-state feedback control.

A key concept in the design of an observer-controller is the *separation principle*, in which the full-state feedback gain $\mathbf{F}$ and the observer gain $\mathbf{L}$ are obtained independently. From these gains, an observer-controller can be constructed, where the observer part provides the state estimate $\hat{\mathbf{x}}(k)$ and the full-state feedback control law is implemented as $u(k) = -\mathbf{F}\hat{\mathbf{x}}(k)$.

Following the controller designs formulated in Chapters 8 and 9, we will perform a control design for the feedback system shown in Figure 10.2, where $\mathbf{r}(k)$ is the reference input. For the single-input, multi-output systems considered in this chapter, $\mathbf{r}(k)$ is a vector having the same dimension as $\mathbf{y}(k)$—namely, $p \times 1$. The $p \times p$ gain matrix $\mathbf{N}$ is introduced as a normalization constant to ensure zero steady-state error for step inputs. This is done because, in a state-space design, the gains are selected only to shift the poles, without regard to the steady-state error.

The state-space equations of the observer-controller, with $\mathbf{D} = \mathbf{0}$, are

$$\hat{\mathbf{x}}(k+1) = (\mathbf{A} - \mathbf{BF} - \mathbf{LC})\hat{\mathbf{x}}(k) - \mathbf{L}[\mathbf{Nr}(k) - \mathbf{y}(k)] \quad (10.5)$$

$$u(k) = -\mathbf{F}\hat{\mathbf{x}}(k) \quad (10.6)$$

where $\hat{\mathbf{x}}$ is the state vector, $\mathbf{Nr}(k) - \mathbf{y}(k)$ is the input to the controller, and $u(k)$ is the scalar output of the controller. Thus in the controller state-space model, the system matrix is $\mathbf{A} - \mathbf{BF} - \mathbf{LC}$, the input matrix is $-\mathbf{L}$, and the output matrix is $-\mathbf{F}$. The closed-loop system in Figure 10.2 is of order $2n$, where n is the order of the plant as well as the observer-controller. The poles of the closed-loop system are the union of the eigenvalues of $\mathbf{A} - \mathbf{BF}$ and $\mathbf{A} - \mathbf{LC}$ (Franklin, Powell, and Emami-Naeini, 1994).

The following example illustrates the design of a controller based on (10.5) and (10.6).

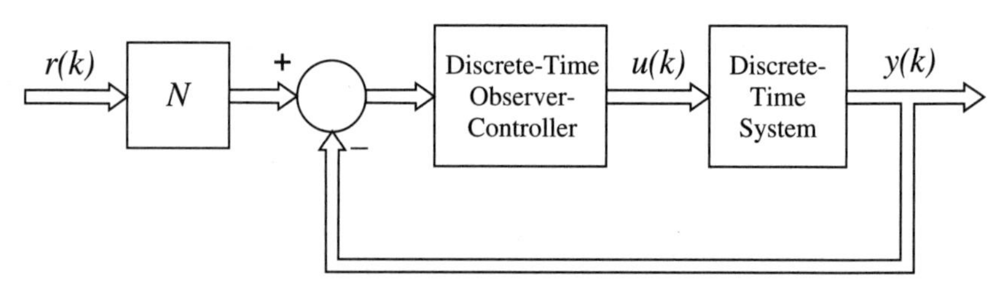

FIGURE 10.2 *Feedback system with observer-controller*

☐ EXAMPLE 10.6 *Observer-Controller Design*

Combine the design results of the pole-placement controller in Example 10.2 and the observer in Example 10.5 to form the observer-controller described by Equations (10.5) and (10.6) for the feedback system in Figure 10.2. Determine the poles and zeros of the transfer function of the observer-controller and plot its frequency response. Form the state-space model of the closed-loop system and compute the closed-loop eigenvalues. Plot the state response of the closed-loop system due to the initial conditions $\mathbf{x}(0) = \begin{bmatrix} 1 & -0.75 & 0.5 \end{bmatrix}^T$ and $\hat{\mathbf{x}}(0) = \mathbf{0}$, with the reference $r(k) = 0$. Then plot the step-reference response of the closed-loop system with zero initial conditions. Assume a sampling time of $T_s = 0.1$ s to perform the simulation.

Solution

We begin, as shown in part (a) of Script 10.6, by using a number of commands from Scripts 10.2 and 10.5 to obtain the full-state feedback gain **F** and the observer gain **L**.

MATLAB Script

```
% Script 10.6: observer-controller design
%-- (a) find control gain F and observer gain L -------------------
Ts = 0.1
A = [0.81   -0.23   -0.045
     0.09    0.98   -0.0023
     0.005   0.10    1     ]            % plant state matrices
B = [0.09 0.0047 0.00016]'
C = [1 3.5 3], D = 0
G = ss(A,B,C,D,Ts)                      % build plant as LTI object
p_s = [0.75 0.80 0.90]'                 % desired system poles
F = place(A,B,p_s)                      % control gain
p_o = [0.50 0.60 0.75]'                 % desired observer poles
L = place(A',C',p_o)'                   % estimator gain
%-- (b) build observer-controller according to (10.5) and (10.6) ---
A_oc = A-B*F-L*C                        % obs-cont system matrix
Goc = ss(A_oc,-L,-F,0,Ts)               % build obs-cont as LTI object
Goc_poles = pole(Goc)
Goc_zeros = tzero(Goc)
bode(Goc)                               % Figure 10.3
Gol = G*Goc                             % controller & plant cascade
Gcl = feedback(Gol,1,-1)                % CL system; DC gain ≠ 1
cl_loop_poles = pole(Gcl)               % CL system poles
%-- (c) pre-gain for unity DC gain --------------------------------
lfg = dcgain(Gcl)                       % low-frequency gain
N = 1/lfg                               % normalization constant
%-- incorporate N in series with closed-loop system transfer function
T_ref = N*Gcl
t = [0:Ts:4];                           % column vector of time
r = 0*t;                                % zero reference input
z0 = [1 -0.75 0.5 0 0 0]'               % initial condition vector
[y,t,z] = lsim(T_ref,r,t,z0);           % IC simulation
% plot system states and observer states
plot(t,z(:,1:3),'o',t,z(:,4),'+',t,z(:,5),'x',t,z(:,6),'^'), grid
legend('x_1','x_2','x_3','x_1hat','x_2hat','x_3hat')
[ys,t,z] = step(T_ref,4);               % compute step resp to 4 sec.
[Mo,tp,tr,ts,ess] = kstats(t,ys,1)      % check time-domain performance
```

From part (b) of Script 10.6, we find the system matrix of the controller to be

$$\mathbf{A} - \mathbf{BF} - \mathbf{LC} = \begin{bmatrix} -1.2701 & -7.0668 & -5.8183 \\ 1.0502 & 4.3638 & 2.9026 \\ -0.8565 & -2.9145 & -1.5837 \end{bmatrix}$$

The controller poles are at $z = -0.1330, 0.8450$, and 0.7980, and the controller zeros are at $z = 0.8559 \pm j0.0594 = 0.8580\epsilon^{\pm j0.0693}$. In general, the controller poles and zeros have no simple relationship to either the desired system poles or the observer poles. The Control System Toolbox has a command, `reg`, that can be used to construct the observer-controller by entering `reg(G,F,L)`. The frequency response of the observer-controller is shown in Figure 10.3.

Denoting the state vector of the closed-loop system as $\mathbf{z}(k)$, which consists of the three states of the plant $\mathbf{x}(k)$ and the three states of the observer $\hat{\mathbf{x}}(k)$, we have

$$\mathbf{z}(k) = \begin{bmatrix} \mathbf{x}(k) \\ \hat{\mathbf{x}}(k) \end{bmatrix}$$

The closed-loop system matrix is

$$\mathbf{A}_{\text{cl}} = \begin{bmatrix} 0.81 & -0.23 & -0.045 & -0.30636 & -0.62882 & -0.45218 \\ 0.09 & 0.98 & -0.0023 & -0.015999 & -0.03284 & -0.023614 \\ 0.005 & 0.1 & 1 & -0.0005446 & -0.00112 & -0.000804 \\ 1.7737 & 6.208 & 5.3211 & -1.2701 & -7.0668 & -5.8183 \\ -0.97618 & -3.4166 & -2.9285 & 1.0502 & 4.3638 & 2.9026 \\ 0.86097 & 3.0134 & 2.5829 & -0.85652 & -2.9145 & -1.5837 \end{bmatrix}$$

which can be written in block-matrix form as

$$\mathbf{A}_{\text{cl}} = \begin{bmatrix} \mathbf{A} & -\mathbf{BF} \\ \mathbf{LC} & \mathbf{A} - \mathbf{BF} - \mathbf{LC} \end{bmatrix}$$

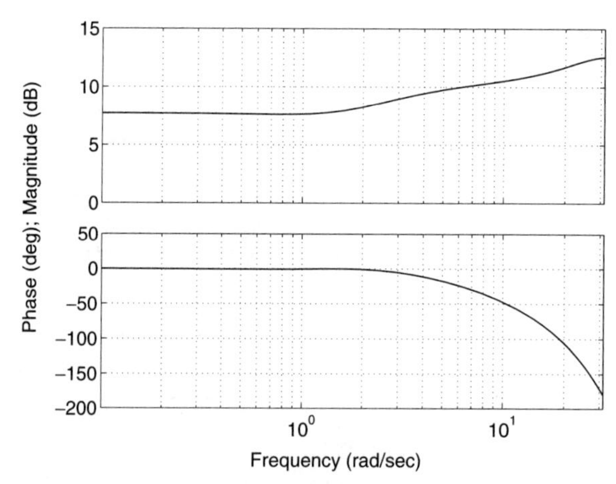

FIGURE 10.3 *Controller frequency response for Example 10.6*

The six eigenvalues of $\mathbf{A}_{\rm cl}$ are $0.50, 0.60, 0.75, 0.75, 0.80$, and 0.90. The first three eigenvalues are due to the observer, and the last three are due to the full-state feedback design, in agreement with the separation principle.

In part (c) of Script 10.6, the normalization constant is found to be $N = 1.0683$. After the gain N is combined with the unnormalized closed-loop system using the `*` operator, we are ready to perform the required simulations.

After assigning the initial conditions in the six-element column vector $z_0 = [1.00\ -0.75\ 0.50\ 0\ 0\ 0]^T$, we use the `lsim` command to generate the six state-variable responses shown in Figure 10.4(a). The initial differences between the system states and the observer states are due to the observer's initial conditions being set to zero.

The step response $y_s(k)$ of the closed-loop system is obtained using the `step` command and is shown in Figure 10.4(b). From the RPI function, `kstats`, we find the step response to have an overshoot of 15.8%, a rise time of 0.23 s, a 2% settling time of 2.5 s, and an error of 0.34% at $t = 4$ s. Because of the normaliztion constant N, this error will go to zero in the steady state.

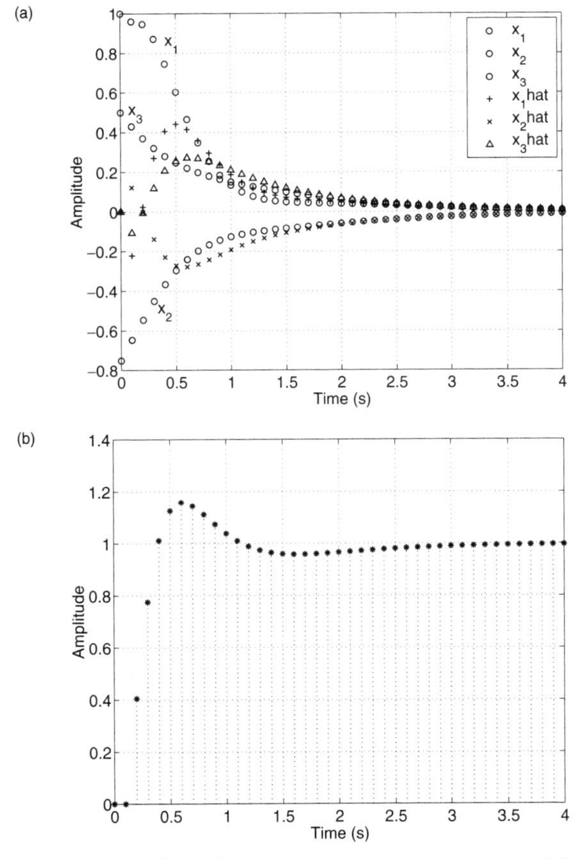

FIGURE 10.4 *Closed-loop system response for Example 10.6: (a) state response $\mathbf{x}(k)$ and $\hat{\mathbf{x}}(k)$; (b) step response $y_s(k)$*

WHAT IF? To investigate the effects on the step response, repeat Example 10.6 using some different values of the full-state feedback and observer eigenvalues. See if you can reduce the 2% settling time. ■

□

REINFORCEMENT PROBLEMS

In each of the following problems, design an observer-controller, following Equation (10.6), for the given state-space model. Use the feedback gain **F** and the observer gain **L** to place the poles of the closed-loop system at the specified locations. Determine the poles and zeros of the controller. Form the closed-loop system model and verify that the closed-loop poles are at the specified locations. Determine the normalization gain **N**, as shown in Figure 10.2, to achieve zero steady-state error for step inputs. Plot the state-variable response of the closed-loop system with $\mathbf{x}(0)$ as given in the problem statement, $\hat{\mathbf{x}}(0) = \mathbf{0}$ and $r(k) = 0$. Then plot the reference-step response of the closed-loop system with zero initial conditions. Assume a sampling time of $T_s = 0.1$ s to perform the simulation.

P10.13 Third-order system.

$$\mathbf{A} = \begin{bmatrix} 0.8238 & 0.0400 & 0.0810 \\ 0.0459 & 0.7799 & 0.1161 \\ 0.0781 & 0.0020 & 0.7448 \end{bmatrix}, \quad \mathbf{B} = \begin{bmatrix} 0.0503 \\ 0.0296 \\ 0.0887 \end{bmatrix}, \quad \text{and} \quad \mathbf{C} = \begin{bmatrix} 0 & 1 & 0 \end{bmatrix}$$

Place the full-state feedback eigenvalues at $z = 0.5$, 0.65, and 0.8, and the observer eigenvalues at $z = 0.3$, 0.4, and 0.6. The initial condition is $\mathbf{x}(0) = \begin{bmatrix} 1 & 0.5 & -0.25 \end{bmatrix}^T$.

P10.14 Fourth-order modal system.

$$\mathbf{A} = \begin{bmatrix} 0.9048 & 0 & 0 & 0 \\ 0 & 0.8187 & 0 & 0 \\ 0 & 0 & 0.7788 & 0 \\ 0 & 0 & 0 & 1 \end{bmatrix}, \quad \mathbf{B} = \begin{bmatrix} 0.0476 \\ 0.0476 \\ 0.0885 \\ 0.5 \end{bmatrix}, \quad \text{and} \quad \mathbf{C} = \begin{bmatrix} 1 & 2 & -1 & 1 \end{bmatrix}$$

Place the full-state feedback eigenvalues at $z = 0.7$, 0.75, 0.85, and 0.95, and the observer eigenvalues at $z = 0.5$, 0.65, 0.78, and 0.92. The initial condition is $\mathbf{x}(0) = \begin{bmatrix} 1 & 2 & 0 & 1 \end{bmatrix}^T$.

P10.15 Lightly damped modes.

$$\mathbf{A} = \begin{bmatrix} 0.9641 & 0.2462 & 0.0303 \\ -0.2462 & 0.9641 & 0.1934 \\ 0 & 0 & 0.6065 \end{bmatrix}, \quad \mathbf{B} = \begin{bmatrix} 0.0056 \\ 0.0528 \\ 0.3935 \end{bmatrix}, \quad \text{and} \quad \mathbf{C} = \begin{bmatrix} 1 & 0 & 0 \end{bmatrix}$$

Place the full-state feedback eigenvalues at $z = -0.9159 \pm j0.2339$ and 0.5762, and the observer eigenvalues at $z = 0.3$, 0.6, and 0.7. The initial condition is $\mathbf{x}(0) = \begin{bmatrix} 1 & 0 & -2 \end{bmatrix}^T$.

EP10.1 Interactive analysis with MATLAB. The steps of the state-space design methods presented in this chapter are programmed in the file `ep10_1.m`. In running the file, the user can enter a state-space model, check its controllability and observability properties, assign the full-state feedback and observer eigenvalues, compute the full-state feedback gain $\mathbf{F}$ and the observer gain $\mathbf{L}$, obtain the observer-controller system matrices and transfer function, calculate the normalization constant $\mathbf{N}$, and simulate the state-variable response due to initial conditions and the output response due to a step-reference input. You can use this file to work on any Reinforcement Problem in this chapter and to experiment with the effect of changing the full-state feedback and observer eigenvalues by merely entering the appropriate system matrices and the desired eigenvalues.

EP10.2 Second-order plant and sensor. Develop an observer-controller for the plant and sensor used in the examples of Chapters 8 and 9 by applying the state-space methods discussed in this chapter. The closed-loop system can be diagrammed showing two outputs, as in Figure 10.5. The transfer functions of the plant and sensor are

$$G_p(s) = \frac{4}{(2s+1)(0.5s+1)} \quad \text{and} \quad H(s) = \frac{1}{0.05s+1}$$

respectively, with a sampling period $T_s = 0.1$ s. The output $y_p(k)$ is that of the plant, and we are interested in its response. The other output, $y_s(k)$, is that of the sensor, and it is this signal that must be used in the observer-controller calculations. Because the sensor's low-frequency gain is unity, the two signals will be identical in the steady-state.

Begin by building state-space models of the plant and sensor and connecting them in series, with the single output being that of the sensor, $y_s(k)$. Test this third-order state-space system for controllability and observability. Then select a set of three closed-loop eigenvalues and do a pole-placement design to obtain the control gain $\mathbf{F}$. Select a set of eigenvalues for the observer and determine the observer gain $\mathbf{L}$.

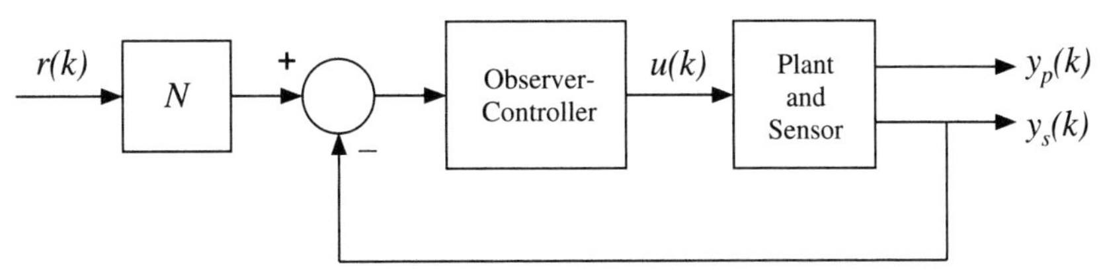

FIGURE 10.5 *Feedback system for Exploratory Problem EP10.2*

Modify your plant/sensor state-space model to provide the output of the plant, $y_P(k)$, and construct the observer-controller. Use the `feedback` command to connect only the sensor output to the feedback summing junction. Also use the `dcgain` command on the closed-loop system to calculate the normalization constant N for the reference input so that the overall system has zero steady-state error for a step input. Then simulate the response of the closed-loop system to a unit-step reference input. Your simulated step response should show both the plant and sensor outputs.

After completing your design, simulate the step response for different combinations of controller and observer eigenvalues. Try to obtain no more than 10% overshoot with a peak time of no more than 0.8 s. To gain more insight into this system, modify your state-space model by adding additional rows to the appropriate $\mathbf{C}$ and $\mathbf{D}$ matrices for the estimated output $\hat{y}(k)$ and the controller output $u(k)$. Having done so, use the `subplot` feature in MATLAB to plot these additional variables and see what you can understand of their behaviors. For example, moving the observer eigenvalues farther from the unit circle in the z-plane should reduce the observer error $y(k) - \hat{y}(k)$. Moving the controller eigenvalues toward the interior of the z-plane should speed up the response but at the expense of larger variations in the controller output $u(k)$, which increases the likelihood of nonlinear behavior due to actuator saturation.

COMPREHENSIVE PROBLEMS

CP10.1 Ball and beam system. Use the file `dbbeam.m` to obtain the discretized state-space model of the ball and beam system using a sampling frequency of 50 Hz. The model has the single input $V_{\text{in}}(k)$, which is the input voltage applied to the servomotor that changes the beam angle. The two outputs $\xi(k)$ and $\theta(k)$ are the ball position and the wheel angle, respectively. First design a full-state feedback control gain $\mathbf{F}$ for the system using the pole-placement technique. In selecting the poles, aim for the two modes associated with the servomotor to decay to 5% of their initial amplitudes within 0.2 s. For the ball-position mode, specify an equivalent pair of continuous-time complex poles that result in a step-response rise time of about 2.5 s and an overshoot of not more than 15%. Then discretize the poles using the transformation $z = \epsilon^{sT_s}$. Obtain the closed-loop system with a full-state feedback and use it to simulate the response for returning the ball to the center from 0.1 m away, with the ball initially at rest and the beam level. Your design should satisfy the requirements that the ball be within 0.01 m of the center in less than 7 s and the wheel angle not exceed ± 1 rad throughout the response.

With the full-state feedback gains obtained above, design an observer-controller for this system using both the ball position and the wheel angle, so the observer gain matrix $\mathbf{L}$ will have four rows and two columns. Design the observer poles so that for the same simulation of returning the ball to the center from 0.1 m away, the ball is still able to settle within 0.01 m in less than 7 s, with the wheel angle not exceeding ± 1 rad. Assume that the initial conditions on the observer states are all zero in the simulation.

CP10.2 Inverted pendulum. Use the file `dstick.m` to obtain the discretized state-space model of the inverted pendulum with a sampling frequency of 100 Hz. The model has the single input $u(k)$, which is the input voltage applied to the DC motor that drives the cart. The two outputs $x(k)$ and $\theta(k)$ are the cart position and the pendulum angle, respectively. First design a full-state feedback control gain $\mathbf{F}$ for the system. The selection of closed-loop system poles should be based on a position step-response rise time of no more than 1.5 s and an overshoot of less than 5%, with the pendulum angle settling to 5% of its initial value in 0.1 s. Simulate this full-state feedback design for the cart returning to the origin from an initial displacement of 0.1 m. You should find that the cart starts in the direction opposite its final position, to get the pendulum falling toward its destination. This observation holds for the subsequent simulations as well.

Because only two states are measured, we need to obtain the cart velocity $v(k)$ and angular acceleration $\alpha(k)$ from $x(k)$ and $\theta(k)$, respectively, using a difference filter for each signal. Implement these filters by starting from a continuous-time highpass filter with the transfer function $100s/(s+100)$. Because two filters are needed, it is necessary to set the poles of the second filter to be slightly different from those of the first to prevent MATLAB from performing a minimal realization. Use the bilinear transform to obtain the corresponding digital filters and implement the full-state feedback controller using these filters. Check the performance of this controller design by performing a simulation using an initial cart displacement of 0.1 m.

Finally, using the full-state feedback gains obtained, design an observer-controller for this system using both outputs, so the observer gain matrix $\mathbf{L}$ will have four rows and two columns. Select the observer poles so that they decay twice as fast as the full-state feedback poles. Simulate the performance of this controller design for the cart returning to the origin from an initial displacement of 0.1 m and compare its performance to that of the other two controllers. Also, use the `impulse` command to simulate the response to a disturbance in the pendulum angle, and use the `step` command to simulate the response to a 0.1 m command in cart position. Comment on the performance of the control system.

CP10.3 Electric power system. Design the voltage regulator in Figure A.8 in Appendix A using the state-space methods discussed in this chapter by proceeding as follows. Run the file `dpower2.m` to obtain the discretized state matrices of the power system sampled at 100 Hz and compute the open-loop poles of the system. The model has seven poles, but only the pole at $z = 0.9989$ is important for voltage regulation. (You can verify this conclusion by observing the open-loop step response.) Thus one design strategy is to use the full-state feedback gain to shift this pole toward the origin of the z-plane. In the observer as well, we need to track only this pole.

a. As a first trial, design the full-state feedback gain $\mathbf{F}$ to shift the open-loop pole at $z = 0.9989$ to $z = 0.98$, and fix the other poles at their open-loop locations. Then design the gain $\mathbf{L}$ so that the observer has a pole at $z = 0.96$ to track the pole at $z = 0.98$, with the other poles at their open-loop locations. Implement the observer-controller according to Equations (10.5) and (10.6). Draw a Bode plot for the controller and simulate the step response of the closed-loop system to a reference voltage command of 0.05. You will find that the controller behaves as a lag controller, and the step response is similar to those found in Comprehensive Problem CP9.2.

b. Now change the full-state feedback pole at $z = 0.98$ and the observer pole at $z = 0.96$ to some other locations, and examine the effect of the variation on the step response. You should see that as they become smaller in magnitude, the rise time becomes faster.

SUMMARY

In this chapter we used MATLAB to determine controllability and observability of discrete-time state-space models, to compute the full-state feedback gain and the observer gain for eigenvalue assignment, and to investigate the performance of the observer-controller. This chapter serves as an introduction to the design of discrete-time controllers using state-space methods. More advanced state-space design algorithms, such as the symmetric root-locus and linear quadratic regulator design, can be found in many control textbooks.

ANSWERS

P10.1 System is controllable.

P10.2 Eigenvalue 0.8187 is not controllable.

P10.3 The diagonal entries of $\mathcal{C}$ are all unity.

P10.4 $\mathbf{F} = \begin{bmatrix} -1.967 & 4.667 & 4.051 \end{bmatrix}$

P10.5 $\mathbf{F} = \begin{bmatrix} 1.597 & 1.130 & -0.281 & 0.295 \end{bmatrix}$

P10.6 (i) $\mathbf{F} = \begin{bmatrix} -0.193 & 0.417 & 0.269 \end{bmatrix}$, $\text{norm}(\mathbf{F}) = 0.533$;
(ii) $\mathbf{F} = \begin{bmatrix} 16.775 & 13.132 & 2.916 \end{bmatrix}$, $\text{norm}(\mathbf{F}) = 21.50$

P10.7 System is observable.

P10.8 Eigenvalue 0.7788 is not observable.

P10.9 The diagonal entries of $\mathcal{O}$ are all unity.

P10.10 $\mathbf{L} = \begin{bmatrix} 4.072 & 1.049 & 1.458 \end{bmatrix}^T$

P10.11 $\mathbf{L} = \begin{bmatrix} 0.189 & -0.169 & 0.0055 & 0.807 \end{bmatrix}^T$

P10.12 $\mathbf{L} = \begin{bmatrix} 0.9347 & 0.4819 & -0.0051 \end{bmatrix}^T$

P10.13 Controller poles: $0.3069\epsilon^{\pm j0.5429}$, and 0.3759; controller zeros: 0.1189 and 0.7018; $N = 1.85$

P10.14 Controller poles: 0.1140, 0.7510, 0.7754, and 0.9573; controller zeros: 0.7525, 0.7799, and 0.9521; $N = 1.0$

P10.15 Controller poles: 0.2818 and $0.6181\epsilon^{\pm j0.2698}$; controller zeros: 0.606 and 7.446; $N = -1.939$

A *Models of Practical Systems*

PREVIEW

In this appendix we present simplified models of two laboratory experiments and two practical plants for which control systems can be developed. These models are used in the comprehensive problems throughout the book to allow the reader to apply MATLAB commands in a quasi-realistic environment (see Table A.1 for a listing of the specific problems). The two practical systems have some interesting behaviors that are generally not encountered in typical textbook problems. The laboratory experiments are commercially available from many vendors. We use model parameters from one of these vendors.

For each of the systems, we provide a linear continuous-time model. For some of the systems, the nonlinear models from which the linear models are derived based on some equilibrium operating conditions are also provided. Then the linear continuous-time model is discretized at an appropriate sampling period using a zero-order hold.

Commands for creating these linear models in both transfer-function and state-space form are contained in M-files available from the Brooks/Cole web site. To load a model, the reader need only enter the name of the appropriate M-file. Both the continuous and discretized models are available.

Keep in mind that a *linear* model has been obtained from a *nonlinear* model at some equilibrium condition. Hence, the linear model always has a limited range of applicability. A bit of common sense, such as making sure that the response range remains inside the linear region, goes a long way in deciding upon reasonable limits of usefulness.

Chapter and topic	*Ball and beam*	*Inverted pendulum*	*Electric power*	*Hydroturbine*
2. Single-block systems	CP2.1	CP2.2	CP2.3	CP2.4
3. Multiblock systems	CP3.1	CP3.2	CP3.3	CP3.4
4. State-space models	CP4.1	CP4.2	CP4.3	CP4.4
5. Sampled-data systems	CP5.4	-	CP5.5	CP5.6
6. Frequency response	-	-	-	-
7. System performance	CP7.1	-	CP7.2	CP7.3
8. Design in z-plane	CP8.1	CP8.2	CP8.3	CP8.4
9. Frequency-response design	CP9.1	-	CP9.2	CP9.3
10. State-space design	CP10.1	CP10.2	CP10.3	-

TABLE A.1 *Listing of comprehensive problems using models presented in Appendix A*

BALL AND BEAM SYSTEM

Figure A.1 shows a ball and beam system in which a DC servomotor is used to rotate a wheel connected to a lever for rolling a metal ball on a beam or track by raising or lowering one end of the track. A schematic of the system is shown in Figure A.2. The wheel angle θ is measured by a potentiometer, and the ball displacement ξ is measured by a resistive strip. This ball and beam system and its variations are used by many universities in their control laboratories.

An analytical model of the experiment can be developed from the dynamic equations governing the servomotor and the ball motion. The electrical part of the servomotor is given by

$$V_{\text{in}} = I_m R + K\omega \tag{A.1}$$

where V_{in} is the input voltage, I_m is the motor current, R is the armature resistance, ω is the wheel velocity, and K is the back emf constant

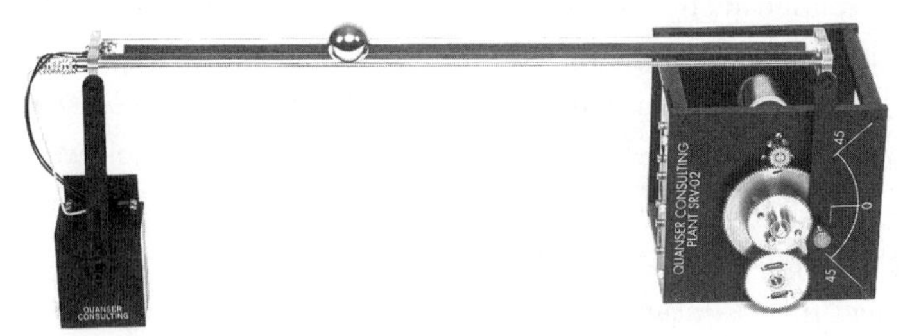

FIGURE A.1 *A ball and beam system (courtesy of Quanser Consulting, Inc.)*

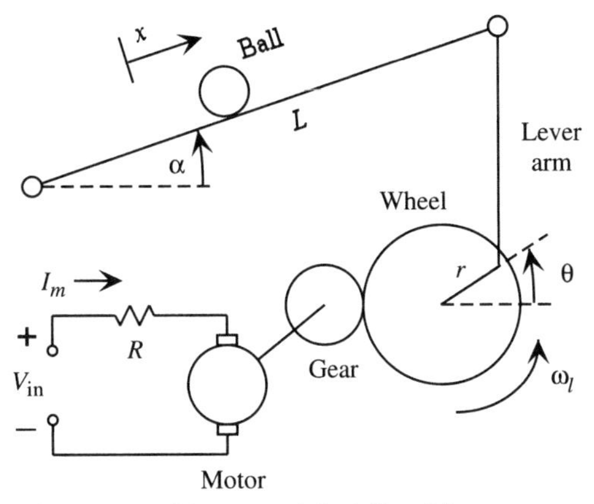

FIGURE A.2 *Schematic of the ball and beam system*

scaled with respect to the angular velocity ω of the wheel. The rotational dynamics of the motor are modeled by

$$\dot{\theta} = \omega \quad \text{and} \quad J_{\text{eq}}\dot{\omega} = KI_m \tag{A.2}$$

where J_{eq}, the equivalent inertia including the motor and wheel inertias, the track, and the ball, is assumed to be constant, and KI_m is the electrical torque. Combining (A.1) and (A.2), we obtain

$$\dot{\theta} = \omega \quad \text{and} \quad \dot{\omega} = -\frac{K^2}{J_{\text{eq}}R}\omega + \frac{K}{J_{\text{eq}}R}V_{\text{in}} \tag{A.3}$$

The rolling ball dynamic equation is given by

$$\ddot{\xi} = -\frac{5}{7}g\sin\alpha - d\dot{\xi} \tag{A.4}$$

where g is the gravitational constant, α is the inclination angle of the track, and d is the friction constant. Most ball and beam models neglect this friction term, which accounts for the ball eventually coming to rest on a level beam after an initial push. Assuming that α is small, we can approximate $\sin\alpha$ by α. Furthermore, from the linkage mechanism shown in Figure A.2, we see that α is a nonlinear function of θ. For small displacements, however, we can approximate α by $(r/L)\theta$. Doing this, we can write (A.4) as

$$\ddot{\xi} = -\frac{5gr}{7L}\theta - d\dot{\xi} \tag{A.5}$$

Taking the Laplace transform of both equations in (A.3) and combining the results, we obtain the transfer function for the servomechanism as

$$\theta(s) = \frac{K/(J_{\text{eq}}R)}{s(s + K^2/(J_{\text{eq}}R))}V_{\text{in}}(s) = G_m(s)V_{\text{in}}(s) \tag{A.6}$$

Similarly, applying the Laplace transform to (A.5), we obtain the transfer function for the ball mechanism as

$$\xi(s) = \frac{-5gr/(7L)}{s(s+d)}\theta(s) = G_b(s)\theta(s) \tag{A.7}$$

Thus the one-input, two-output model of the ball and beam system in transfer function form is given by

$$\begin{bmatrix} \xi(s) \\ \theta(s) \end{bmatrix} = \begin{bmatrix} G_b(s)G_m(s) \\ G_m(s) \end{bmatrix} V_{\text{in}}(s) \tag{A.8}$$

where

$$G_b(s) = \frac{-5gr/(7L)}{s(s+d)} \quad \text{and} \quad G_m(s) = \frac{K/(J_{\text{eq}}R)}{s(s+K^2/(J_{\text{eq}}R))} \tag{A.9}$$

To obtain the state-space model of the system, we directly combine (A.3) and (A.5) to obtain

$$\dot{\mathbf{x}} = \mathbf{A}\mathbf{x} + \mathbf{B}u, \quad \mathbf{y} = \mathbf{C}\mathbf{x} \tag{A.10}$$

where

$$\mathbf{x} = \begin{bmatrix} \xi \\ \dot{\xi} \\ \theta \\ \omega \end{bmatrix}, \quad u = V_{\text{in}}, \quad \mathbf{y} = \begin{bmatrix} \xi \\ \theta \end{bmatrix}$$

$$\mathbf{A} = \begin{bmatrix} 0 & 1 & 0 & 0 \\ 0 & -d & \dfrac{5gr}{7L} & 0 \\ 0 & 0 & 0 & 1 \\ 0 & 0 & 0 & -\dfrac{K^2}{J_{\text{eq}}R} \end{bmatrix}, \quad \mathbf{B} = \begin{bmatrix} 0 \\ 0 \\ 0 \\ \dfrac{K}{J_{\text{eq}}R} \end{bmatrix}, \quad \text{and}$$

$$\mathbf{C} = \begin{bmatrix} 1 & 0 & 0 & 0 \\ 0 & 0 & 1 & 0 \end{bmatrix}$$

The model parameter values given in the experiment manual provided by the manufacturer (Apkarian, 1997) are $K = 0.54$ V/(rad/s), $J_{\text{eq}} = 0.00291$ kg-m^2, $R = 2.6\ \Omega$, $g = 9.8$ m/s^2, $r = 0.0125$ m, and $L = 0.4318$ m. The nominal value of the coefficient d measured experimentally is $0.15\ \text{s}^{-1}$. The input and output variables have the following units: meters for ξ, radians for θ, and volts for V_{in}. Because of the configuration of the lever, θ is limited to be within $\pm\pi/2$ rad.

The continuous-time and discretized models of the system are available from the file `dbbeam.m`. The sampling frequency used in `dbbeam.m` for the discretization is 50 Hz.

There are several possible control structures that can be applied to this system, which essentially consists of two subsystems in cascade. For lead-lag control design, the feedback structure in Figure A.3 can be used.

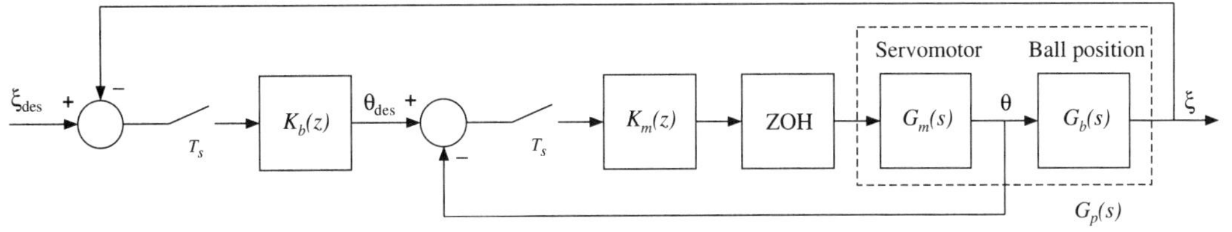

FIGURE A.3 *Ball and beam system with two controllers*

The controller $K_m(z)$ closes the servomechanism loop to regulate the beam angle, and the controller $K_b(z)$ regulates the ball position. This control structure is investigated in several comprehensive problems. For a state-space design, observers can be used to reconstruct the states in conjunction with a full-state feedback gain.

INVERTED PENDULUM

Figure A.4 shows a cart on wheels that can be moved back and forth to balance a pendulum that is supported by a frictionless pivot at the base of the stick. A schematic of the system is shown in Figure A.5. The wheel is driven by a DC servomotor and is geared to move along the track without slippage. A potentiometer measures the position of the cart and another potentiometer measures the angle made by the pendulum with the vertical. Like the ball and beam system, this inverted pendulum system and its variations are commonly found in university control laboratories. To develop a controller for the system, we want to have a linear mathematical model in state-space form.

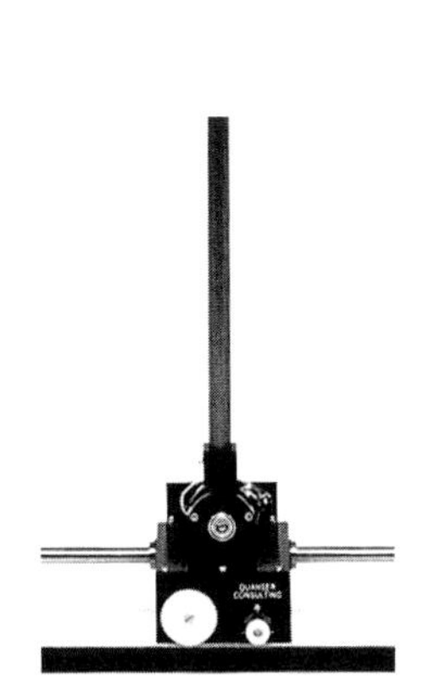

FIGURE A.4 *Cart supporting a pendulum (courtesy of Quanser Consulting, Inc.)*

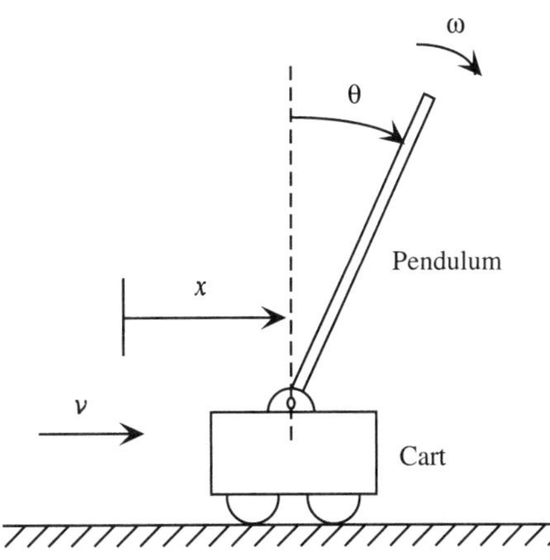

FIGURE A.5 *Positive senses of the variables describing the motion of the inverted pendulum system*

Such a model can be developed by drawing free-body diagrams of the stick and cart and the circuit diagram for the electric motor. From the free-body diagrams, the forces and moments can be summed to obtain the equations of motion. The nonlinear terms can be neglected, provided that the angle of the stick with the vertical remains sufficiently small (say, no more than $\pm 10°$). The electrical equation relates the input voltage of the motor to the translational force applied to the cart.

The variables used to describe the system are θ, the angle of the pendulum relative to the vertical, in radians; ω, the derivative of that angle, in rad/s; x, the horizontal position of the cart, in meters; v, the velocity of the cart, in m/s; and u, the control signal applied to the armature of the motor, in volts. The positive senses of these variables are shown in Figure A.5. The interested reader can consult Franklin, Powell, and Emami-Naeini (1994) for a derivation of the model.

The free-body diagram yields the following set of coupled differential equations:

$$\begin{aligned} (m_p + m_c)\dot{v} + m_p\dot{\omega}I_p\cos\theta - m_p\omega^2 I_p\sin\theta &= F \quad \text{(A.11)} \\ m_p I_p\dot{v}\cos\theta - m_p I_p\omega v\sin\theta + m_p I_p^2\dot{\omega} - m_p g I_p\sin\theta &= 0 \quad \text{(A.12)} \end{aligned}$$

where m_p is the mass of the pendulum, m_c is the mass of the cart, I_p is the center of gravity of the pendulum, and F is the applied force. The electrical equation of the motor is given by

$$F = \frac{K}{R}u - \frac{K^2}{R}v \tag{A.13}$$

where R is the armature resistance and K is the back emf constant scaled to v.

Linearizing (A.11) through (A.13) and selecting the state vector to be $[\theta\ \omega\ x\ v]^T$, we obtain the state-space model as

$$\frac{d}{dt}\begin{bmatrix}\theta \\ \omega \\ x \\ v\end{bmatrix} = \begin{bmatrix} 0 & 1 & 0 & 0 \\ \dfrac{(m_p+m_c)g}{m_c I_p} & 0 & 0 & \dfrac{K^2}{m_c I_p R} \\ 0 & 0 & 0 & 1 \\ -\dfrac{m_p g}{m_c} & 0 & 0 & -\dfrac{K^2}{m_c R} \end{bmatrix}\begin{bmatrix}\theta \\ \omega \\ x \\ v\end{bmatrix} + \begin{bmatrix} 0 \\ -\dfrac{K}{m_c I_p R} \\ 0 \\ \dfrac{K}{m_c R}\end{bmatrix}u$$

For control purposes, we assume that there are only two measurable output variables. One of these is the cart position x, obtained from a potentiometer connected to the wheels. The other signal, θ, is obtained from a potentiometer mounted on the cart and connected to the base of the stick.

The file `dstick.m` develops the model when it is assumed to be in either the TF or the SS form. The values of the model parameters given in the experiment manual (Apkarian, 1997) are $m_p = 0.21$ kg, $m_c = 0.455$ kg, $I_p = 0.305$ m, $R = 2.6\ \Omega$, $K = 4.47$ N/A, and $g = 9.8$ m/s^2. For the discretized model, a sampling frequency of 100 Hz is used.

Several control structures can be applied to the inverted pendulum system. One approach is to design a full-state feedback control law and implement it using an observer. Another approach, shown in Figure A.6, involves designing the controller $K_p(z)$ to stabilize the pendulum and the controller $K_c(z)$ to control the cart position. This control structure is investigated in Comprehensive Problem CP8.2.

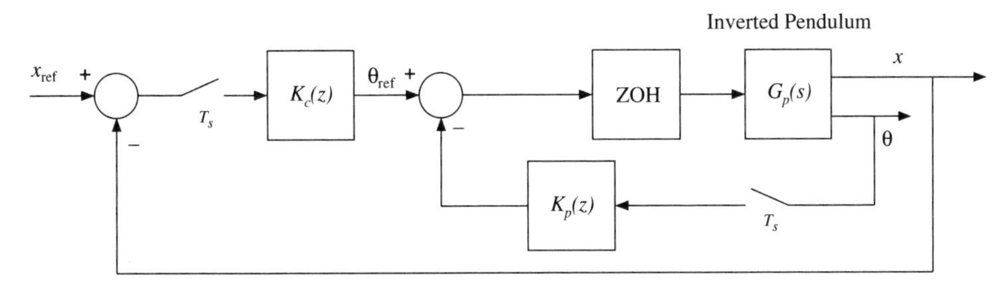

FIGURE A.6 *Inverted pendulum system with two controllers*

ELECTRIC POWER SYSTEM

The single-machine infinite-bus system in Figure A.7 is usually used as the first step in designing an excitation system controller for a generator delivering the electrical power P. The objective is to design a feedback controller with output $u(t)$ to regulate the field voltage such that the machine terminal voltage V_{term} is maintained at a desired value V_{ref}. In this example, the machine model includes higher-order flux dynamics, and the field voltage actuator is a solid-state rectifier. The states for the machines are the rotor angle δ and speed ω, and the direct- and quadrature-axis fluxes E_q', ψ_{1d}, E_d', and ψ_{1q}. The rotor angle δ is in radians; the speed ω is in per unit (pu), normalized with respect to the synchronous speed; and the fluxes are also in pu, normalized with respect to the rated voltage. The actuator is modeled with the state V_R.

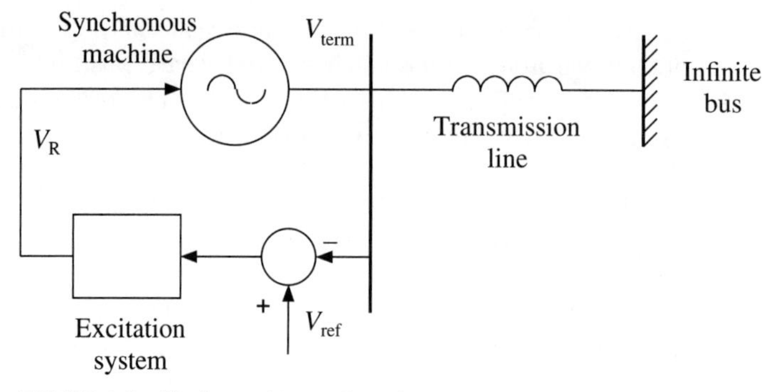

FIGURE A.7 *Single-machine infinite-bus system*

The linearized system model is given by

$$\dot{\mathbf{x}} = \mathbf{A}\mathbf{x} + \mathbf{B}u(t) \quad \text{and} \quad \mathbf{y} = \mathbf{C}\mathbf{x}$$

where

$$\mathbf{x} = \begin{bmatrix} \delta & \omega & E_q' & \psi_{1d} & E_d' & \psi_{1q} & V_R \end{bmatrix}^T$$

$$\mathbf{y} = \begin{bmatrix} V_{\text{term}} & \omega & P \end{bmatrix}^T$$

For a particular operating condition, the numerical values of the state-space matrices are

$$\mathbf{A} = \begin{bmatrix} 0 & 377.0 & 0 & 0 & 0 & 0 & 0 \\ -0.246 & -0.156 & -0.137 & -0.123 & -0.0124 & -0.0546 & 0 \\ 0.109 & 0.262 & -2.17 & 2.30 & -0.0171 & -0.0753 & 1.27 \\ -4.58 & 0 & 30.0 & -34.3 & 0 & 0 & 0 \\ -0.161 & 0 & 0 & 0 & -8.44 & 6.33 & 0 \\ -1.70 & 0 & 0 & 0 & 15.2 & -21.5 & 0 \\ -33.9 & -23.1 & 6.86 & -59.5 & 1.50 & 6.63 & -114.0 \end{bmatrix}$$

$$\mathbf{B} = \begin{bmatrix} 0 & 0 & 0 & 0 & 0 & 0 & 17.6 \end{bmatrix}^T$$

$$\mathbf{C} = \begin{bmatrix} -0.123 & 1.05 & 0.230 & 0.207 & -0.105 & -0.460 & 0 \\ 0 & 1 & 0 & 0 & 0 & 0 & 0 \\ 1.42 & 0.900 & 0.787 & 0.708 & 0.0713 & 0.314 & 0 \end{bmatrix}$$

The dominant mode of the system is a pair of lightly damped poles (referred to as the swing mode) corresponding to the mechanical oscillations of the machine versus the infinite bus.

This system is used in several comprehensive problems for the design of a digital voltage control system having the structure shown in Figure A.8, where $K_V(z)$ is the transfer function of the controller. The terminal voltage V_{term} is to be regulated with respect to the reference voltage V_{ref}. The dynamics of the sensor for measuring the terminal voltage are neglected. As seen in the comprehensive problems, the gain of the voltage regulator is limited by the lightly damped mode. In practice, a second feedback loop using either the speed or power output is applied to improve the damping on this lightly damped mode. In this book, we simplify the process by adding some mechanical damping to this mode in the system model.

The model of the power system is contained in the file `dpower.m`, and the model with additional damping is in the file `dpower2.m`. Because this is a multi-output system, the user should select whichever output is to be considered. For a state-space model, the appropriate row of the **C** matrix should be used. In `dpower.m` the continuous-time model is discretized with a sampling rate of 100 Hz, which is consistent with industry practice for the sampling rate of a digital voltage regulator.

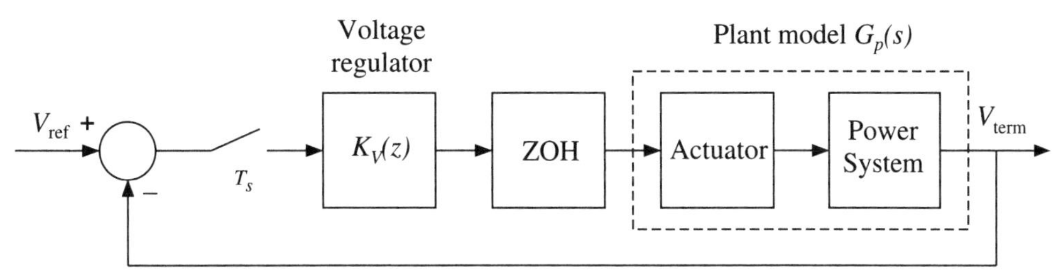

FIGURE A.8 *Excitation control system*

HYDROTURBINE AND PENSTOCK

Figure A.9 shows a simplified sketch of a hydroturbine and penstock system. The nonlinear power generation equations are

$$H = v^2/Q^2, \quad \dot{v} = (g/L)(H_o - H), \quad \text{and} \quad P = avH$$

where H is the pressure at the gate, v the water velocity, Q the effective gate opening, g the gravity constant, L the length of the penstock, H_o the reservoir water pressure, P the mechanical power generated by the turbine, and a the area of the penstock. The units of these variables are not needed in this problem. Because the power P is proportional to the water velocity v, to increase P one has to increase the gate opening Q

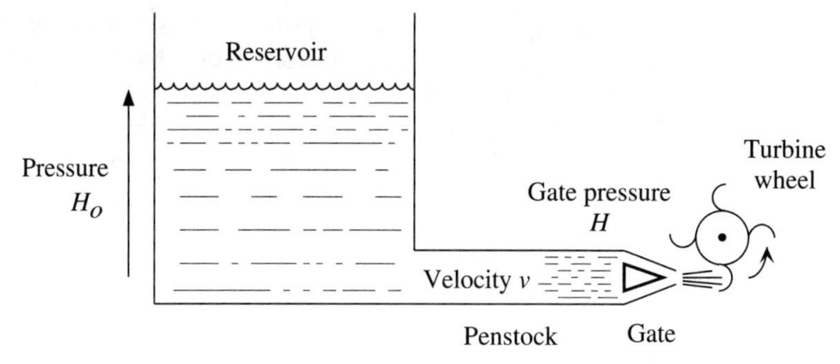

FIGURE A.9 *A hydroturbine and penstock system*

such that v increases. The water velocity in the penstock does not change instantaneously, however, and if Q is increased, H will momentarily decrease. As a result, P first decreases until the water velocity increases sufficiently. Such systems exhibiting that *it will get worse before it gets better* are quite common in engineering and economic systems.

Assuming that the nonlinear model is in a condition of equilibrium with $H = H_o$ and $v = v_o$, we obtain the linearized model of the hydroturbine as the transfer function

$$\Delta P(s) = \left(\frac{1 - sT_w}{1 + sT_w/2}\right)\Delta Q(s)$$

where $\Delta P(s)$ and $\Delta Q(s)$ are the Laplace transforms of the incremental power and gate opening, respectively, and $T_w = (v_o l)/(H_o g)$ is the water time constant. The system has a right half-plane zero at $s = 1/T_w$.

To complete the model, we represent the dynamics of the movement of the gate opening with an actuator by

$$\Delta Q(s) = \left(\frac{1}{T_Q s + 1}\right)\Delta U(s)$$

where T_Q is the actuator time constant and $\Delta U(s)$ is the transform of the incremental actuation signal generated by the controller. Typical time constants are $T_w = 2$ s and $T_Q = 0.5$ s.

The model and data for the hydroturbine system are contained in the file `dhydro.m`. From the time response of the continuous-time system, we determine that a sampling rate of 10 Hz is adequate for this system. The discretized system is used in several comprehensive problems to analyze the effect on the system response of a zero outside the unit circle. It is also used to study the generation control system in Figure A.10. For notational simplicity, we will drop the Δ's from the incremental variables. In this feedback system, the controller $G_c(z)$ utilizes the difference between the desired power P_{des} and the actual power P to generate an output $u(k)$ to control the gate position.

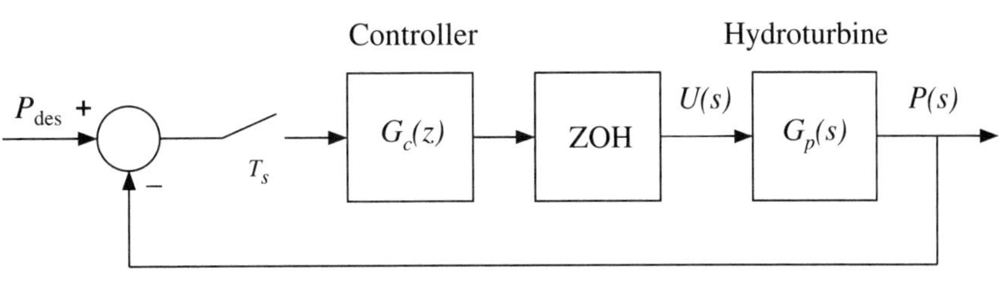

FIGURE A.10 *Block diagram of the hydroturbine and penstock control system*

B *Discrete Fourier Series and Transform*

PREVIEW

In the context of signal processing and filtering design, it is useful to express a discrete-time signal $x(k)$, sampled uniformly with a period of T_s, in the frequency domain (Oppenheim and Schafer, 1999). In this appendix, we summarize the main tools for expressing discrete-time signals in the frequency domain: discrete Fourier series for periodic discrete-time signals and discrete Fourier transform for finite-duration discrete-time signals. These techniques have important practical applications, including digital filtering and system identification.

DISCRETE FOURIER SERIES

Consider first a periodic discrete-time signal $\tilde{x}(k)$ with a positive integer N as its period. It follows that $\tilde{x}(k)$ satisfies the periodicity property

$$\tilde{x}(k+N) = \tilde{x}(k), \quad \text{for all } k \tag{B.1}$$

We define a set of N fundamental periodic discrete-time complex exponential signals, with period N, as

$$\phi_\ell(k) = \epsilon^{j\ell(2\pi/N)k}, \quad \ell = 0, 1, \ldots, N-1 \tag{B.2}$$

which are vectors of unit length spaced $(2\pi/N)k$ rad apart. Then the periodic signal $\tilde{x}(k)$ can be expressed as the discrete-time Fourier series

$$\tilde{x}(k) = \sum_{\ell=0}^{N-1} \tilde{a}_\ell \phi_l(k), \tag{B.3}$$

where the Fourier series coefficients $\tilde{a}_\ell$ are given by

$$\tilde{a}_\ell = \frac{1}{N} \sum_{k=0}^{N-1} \tilde{x}(k)\phi_k(-\ell), \quad \ell = 0, 1, \ldots, N-1 \tag{B.4}$$

Equations (B.3) and (B.4) are known as the *synthesis* and *analysis* equations, respectively, of the discrete-time Fourier series.

DISCRETE FOURIER TRANSFORM

For a finite-duration discrete-time signal $x(k)$, $k = 0, 1, \ldots, N-1$, consisting of N data points, we can apply the discrete Fourier transform to compute its spectrum. The discrete Fourier transform of the signal $x(k)$ is given by

$$X(\ell) = \sum_{k=0}^{N-1} x(k) W_N^{k\ell}, \quad \ell = 0, 1, \ldots, N-1 \tag{B.5}$$

where $W_N = \epsilon^{-j2\pi/N}$ is the fundamental frequency phasor of unit length with a phase of $-2\pi/N$. The frequency coefficient $X(\ell)$ is in general complex and represents the amplitude and phase of $x(k)$ at the discrete frequency point $(\ell/N)f_s$, where $f_s = 2\pi/T_s$ is the sampling frequency.

Given the frequency coefficients, the time-domain signal can be obtained from the expression

$$x(k) = \frac{1}{N} \sum_{\ell=0}^{N-1} X(\ell) W_N^{-k\ell}, \quad k = 0, 1, \ldots, N-1 \tag{B.6}$$

Equations (B.5) and (B.6) are known as the Fourier transform *analysis* and *synthesis* equations, respectively.

Directly computing the expressions in (B.5) requires N^2 multiply and add operations. Because of the circular properties of the exponents $W_N^{k\ell}$, however, the amount of computation can be reduced to $N \log_2(N)$ operations if N is a power of 2. This reduction is particularly significant if N is large. In fact, the number of operations can be reduced in all cases where N is not a prime number. Such algorithms are known as the *fast Fourier transform*. The MATLAB `fft` function performs the fast Fourier transform computation for (B.5), and the inverse function `ifft` computes the time-domain signal $x(k)$ (B.6) from the frequency coefficients.

For a periodic signal $\tilde{x}(k)$ of period N, the `fft` function can be used to compute its Fourier series coefficients by noting that

$$\tilde{a}_\ell = X(\ell)/N, \quad \ell = 0, 1, \ldots, N-1 \tag{B.7}$$

The following example shows the application of the `fft` command to a sinusoidal signal.

Consider a discrete-time sinusoidal signal $x(k) = 3\cos(20\pi k T_s + 30°)$ with a frequency of 10 Hz, where $T_s = 0.01$ s. The period of the signal is $N = 10$. The commands in Script B.1 perform the `fft` computation.

MATLAB Script

```
% Script B.1:  computation of the spectrum of a discrete-time signal
Ts = 0.01                               % sampling period
N = 10                                  % period
k = [0:1:N-1]'                          % time index
x = 3*cos(20*pi*k*Ts+30*pi/180)         % generate signal
X = fft(x)                              % discrete Fourier transform
a = X/N                                 % Fourier series coefficients
[a_mag,a_ph] = xy2p(a)                  % convert to polar form
a_ph_deg = a_ph*180/pi                  % convert phase to degrees
```

The result shows that the 10 elements of the discrete Fourier transform are

$$X = [0 \;\; 12.99 + j7.5 \;\; 0 \;\; 0 \;\; 0 \;\; 0 \;\; 0 \;\; 0 \;\; 0 \;\; 12.99 - j7.5]$$

The Fourier series coefficients are obtained by dividing X by $N = 10$, yielding

$$\tilde{a} = [0 \;\; 1.5\epsilon^{j\pi/6} \;\; 0 \;\; 0 \;\; 0 \;\; 0 \;\; 0 \;\; 0 \;\; 0 \;\; 1.5\epsilon^{-j\pi/6}]$$

The nonzero entries $\tilde{a}_1 = 1.5\epsilon^{j\pi/6}$ and $\tilde{a}_9 = 1.5\epsilon^{-j\pi/6}$ correspond to the Fourier series coefficients with frequencies $(1/10)100 = 10$ Hz and $(9/10)100 = 90$ Hz, respectively. With the sampling frequency of 100 Hz, however, the 90 Hz can be considered as a frequency of -10 Hz. Combining the positive and negative frequencies, we obtain a cosine signal of 10 Hz with an amplitude of 3 (double the amplitude of the individual Fourier series coefficients) and a phase of 30°.

C Root-Locus Plots

PREVIEW

Root-locus plots in the s-plane have traditionally been used for the design of continuous-time linear feedback systems. The technique is equally applicable to discrete-time linear feedback systems when the plots are drawn in the z-plane. In this appendix we illustrate the use of MATLAB commands to make root-locus plots.

If the open-loop transfer function $F(z)$ of an nth-order single-input/single-output discrete-time system is available as the LTI object `F` and contains a loop gain K, the root-locus plot for the closed-loop system in Figure C.1 can be generated by entering the `rlocus(F)` command. The resulting plot shows how the closed-loop poles move in the complex z-plane as the loop gain K varies from 0 to ∞. Such a plot is known as the 180°-locus. MATLAB computes the n branches of the root locus and draws the plot, with the real and imaginary axes it selects. The numerical values of the roots corresponding to the values of K that MATLAB selects are not available for viewing or printing, however.

Alternatively, one can use separate commands to generate a set of gains, calculate the roots, and plot them. In either case, grid lines and a title must be added, and the user has the option of adjusting the plotting region by using MATLAB's `axis` command. If the `axis equal` command is given, the same scaling will be used for the real and the imaginary axes, thereby representing angles correctly and avoiding distortion of the locus. A circle will appear as a circle, for example. For the `axis` command to take effect on a root-locus plot, it needs to be entered after the `ucircle` command but before the `rlocus` command, as shown in Script C.1 for the example that follows.

Once the locus has been drawn, the `rlocfind` command can be used interactively to determine the value of K corresponding to any point on the locus that the user selects with a mouse click. The values of all n closed-loop poles corresponding to this value of K can be displayed by using the command in the form `[K,pCL] = rlocfind(F)`. Alternatively, the Control System Toolbox data marker feature can be used to find the value of K for a particular pole location (see Example 8.1).

The 0°-locus corresponds to either a positive sign where the feedback path enters the feedback summing junction or negative values for the loop gain K. To have MATLAB draw the 0°-locus, we can insert a negative sign in front of the system object and proceed as before, with the command `rlocus(-F)`.

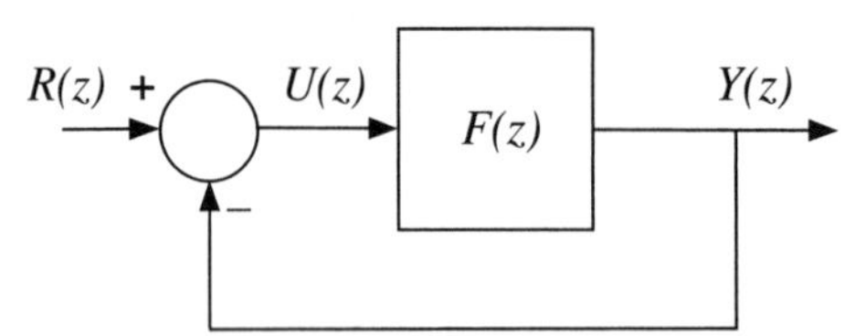

FIGURE C.1 *Feedback system for root-locus analysis*

To illustrate the use of these commands, we use MATLAB to draw the root locus for the discrete-time open-loop transfer function

$$F(z) = \frac{K(z^2 + 1.1z + 0.24)(z - 0.7)}{(z^2 - 0.08z - 0.882)(z^2 - 1.8z + 0.833)}$$

where the sampling period is unity. Then we use the `rlocfind` command to determine the values of K for which (i) the two complex branches move outside the unit circle, (ii) the two complex branches return inside the unit circle, and (iii) the leftmost real branch crosses outside the unit circle. Based on these results, we establish the range of nonnegative loop gain K for which the closed-loop system is stable.

The commands in MATLAB Script C.1 build the open-loop transfer function (with $K = 1$) in terms of the numerator and denominator polynomials and then compute and plot the root locus. The zeros and poles of the open-loop transfer function $F(z)$ are given in Table C.1, and the root-locus plot produced by MATLAB is shown in Figure C.2.

MATLAB Script

```
% Script C.1:  Root locus for 4th-order system with real & complex
%              open-loop poles
Ts = 1                                             % sampling period
num = conv([1 1.1 0.24],[1 -0.7])                  % 3rd-order numerator
den = conv([1 -0.08 -0.882],[1 -1.8 0.833])  % 4th-order denominator
F = tf(num,den,Ts)                                 % OL sytem as TF object
[zF,pF,kF] = zpkdata(F,'v')                        % OL zeros, poles, & gain
ucircle, hold on                                   % unit circle
axis equal                                         % equal scaling
rlocus(F), hold off                                % compute and plot locus
[kk,clroots] = rlocfind(F)                         % calculate gain & CL poles
```

Zeros:	$z_1 = -0.80$	$z_2 = 0.70$	$z_3 = -0.30$	
Poles:	$p_1 = 0.9127\epsilon^{j0.1669}$	$p_2 = 0.9127\epsilon^{-j0.1669}$	$p_3 = -0.90$	$p_4 = 0.98$

TABLE C.1 *Open-loop zeros and poles of* $F(z)$

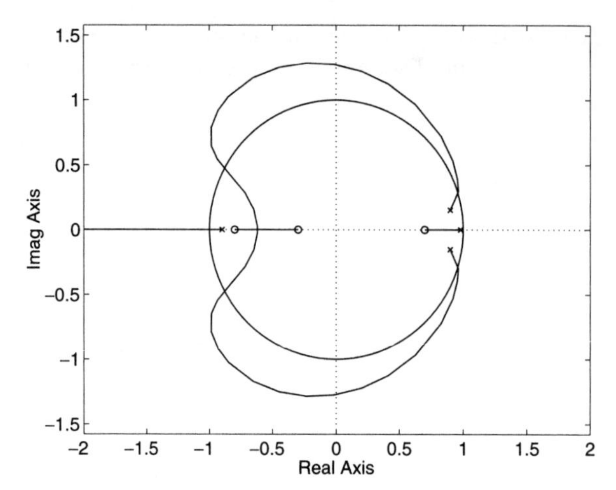

FIGURE C.2 *Root locus produced by Script C.1*

We see that the real-axis portions of the root locus are (i) $\text{Re}(z) \leq -0.90$, (ii) $-0.80 \leq \text{Re}(z) \leq -0.30$, and (iii) $0.70 \leq \text{Re}(z) \leq 0.98$. There are four branches, one starting from each of the open-loop poles at $z = -0.90, 0.9127\epsilon^{\pm j0.1669}$, and 0.98. Three branches terminate on the open-loop zeros at $z = -0.80, -0.30$, and 0.70, and the other branch goes to infinity as $K \to \infty$, along the negative real axis.

Using the `rlocfind` command, we find that the two branches starting from the complex open-loop poles p_1 and p_2 at $z = 0.9127\epsilon^{\pm 0.1669}$ move outside the unit circle at $z = 1.0\epsilon^{\pm j0.2809}$ for $K = 0.0354$ and reenter the unit circle at $z = 1.0\epsilon^{\pm j2.647}$ for $K = 5.24$. These two branches meet one another on the real axis at $z = -0.619$ and then terminate on the zeros at $z_1 = -0.80$ and $z_3 = -0.30$. The branch starting at $p_3 = -0.90$ leaves the unit circle at $z = -1$ for $K = 3.03$. Finally, the branch that leaves $p_4 = 0.98$ and terminates on $z_2 = 0.70$ remains within the unit circle for all positive K. Therefore the only range of positive gains for which *all four* of the branches are within the unit circle is $0 < K < 0.0354$.

REINFORCEMENT PROBLEMS

Use MATLAB to draw root-locus plots for the open-loop transfer functions given in the following statements. In each case, draw an initial plot without specifying the boundaries of the plotting area. Then redo the plot using the `axis` command with appropriate arguments to best show the region of interest. After the plot has been drawn, use the `rlocfind` command to determine the values of K that correspond to selected points on the locus, such as where it crosses the unit circle. In each case, use a sampling period of unity.

PC.1 Two real poles and one zero.

$$F(z) = \frac{K(z-0.1)}{(z-0.75)(z-0.9)}$$

Because the number of poles exceeds the number of zeros by only one, only one branch goes to infinity as $K \to \infty$.

PC.2 Four poles.

$$F(z) = \frac{K(z-0.1)}{(z+0.6)(z+0.3)(z-0.75)(z-0.5)}$$

Because $F(z)$ has four poles and one zero, there are three branches that approach infinity as $K \to \infty$.

PC.3 Complex poles.

$$F(z) = \frac{K(z+0.5)}{(z-0.6)(z^2-0.4z+0.29)}$$

This transfer function has a pair of complex poles in addition to a single real pole and a single real zero.

PC.4 Complex zeros.

$$F(z) = \frac{K(z^2+1.2z+0.85)}{(z+0.1)(z-0.8)}$$

This transfer function has a pair of complex zeros and two real poles. Because the number of zeros equals the number of poles, no branches will approach infinity as K becomes large.

PC.5 Pole outside the unit circle.

$$F(z) = \frac{K(z+0.5)(z-0.6)}{(z-0.13)(z-0.5)(z-1.2)}$$

This transfer function has a pole outside the unit circle in the z-plane, which means that the open-loop system is unstable. The closed-loop system is unstable for low values of K because the branch that starts from the pole at $z = 1.2$ is outside the unit circle. Once this branch crosses the unit circle at $z = 1$, the closed-loop system will be stable because all three branches will be inside the unit circle.

PC.6 Double pole at $z = 1$.

$$F(z) = \frac{K(z-0.37)(z-0.6)}{(z-1)^2(z+0.8)(z+0.4)}$$

The locus has four branches, with two of them emanating from the double pole at $z = 1$. Two branches approach infinity as $K \to \infty$.

PC.7 Triple pole at $z = 1$.

$$F(z) = \frac{K(z-0.6)(z-0.8)}{(z-1)^3}$$

This transfer function has a triple pole at $z = 1$ and two real zeros. There are three branches starting from the triple pole, with two of them starting outside the unit circle and then being pulled back into the interior of the unit circle by the presence of the two zeros. One branch goes to infinity as $K \to \infty$, and the other two branches terminate on the two zeros.

PC.8 Negative gain.

$$F(z) = \frac{K(z-0.4)}{(z-0.75)(z-0.9)(z+0.3)(z-0.1)} \quad \text{where } K < 0$$

To account for the fact that $K < 0$, we can insert a negative sign in front of the numerator and draw the locus for $K > 0$. Find the value of K where the locus crosses the unit circle.

PC.9 State-space model. Use the MATLAB `rlocus` command to make a root-locus plot of the state-space model for which

$$\mathbf{A} = \begin{bmatrix} 0.15 & 0.5 & 0.1 \\ 0 & 1.6 & -0.68 \\ 0 & 1 & 0 \end{bmatrix}, \quad \mathbf{B} = \begin{bmatrix} 0 \\ 0.45 \\ 0 \end{bmatrix}$$

$$\mathbf{C} = \begin{bmatrix} 0.9 & 0 & 0 \end{bmatrix}, \quad \mathbf{D} = \begin{bmatrix} 0 \end{bmatrix}$$

This root locus is the variation of the closed-loop system poles with the feedback control $u = Ky$ for $0 \le K < \infty$.

ANSWERS

PC.1 One branch leaves the unit circle at $z = -1$ for $K = 3.02$; for this gain, the other closed-loop pole is $z = -0.374$.

PC.2 For $K = 0.675$, one branch leaves the unit circle at $z = -1$; for this gain, the other two closed-loop poles are at $z = 0.884\epsilon^{\pm j0.701}$.

PC.3 For $K = 0.427$, two complex-conjugate branches leave the unit circle at $z = 1.0\epsilon^{\pm j1.024}$; for this gain, the other closed-loop pole is at $z = -0.04$.

PC.4 There are two branches, both of which remain inside the unit circle for all $K > 0$. These branches meet on the real axis at $z = 0.22$ for $K = 0.160$.

PC.5 One branch starts outside the unit circle and crosses to its interior at $z = 1$ for $K = 0.143$. The two complex branches leave the unit circle at $z = 1.0\epsilon^{\pm j1.873}$ for $K = 1.80$; for this value of gain, the other closed-loop pole is at $z = 0.616$.

PC.6 Two branches start at $z = 1$ and remain inside the unit circle for all $K > 0$. Two complex branches leave the unit circle at $z = 1.0\epsilon^{\pm j1.95}$ for $K = 1.63$.

PC.7 Three branches start at the triple open-loop pole at $z = 1$. One of these branches goes along the real axis, inside the unit circle. The other two branches leave the triple pole at angles of $\pm 60°$, moving outside the unit circle. They reenter it at $z = 1.0\epsilon^{\pm j0.405}$ for $K = 0.312$. These two branches meet on the real axis at $z = 0.05$ for $K = 2.08$. One of these branches leaves the unit circle at $z = -1$ for $K = 2.78$.

PC.8 There are four branches. One leaves the unit circle at $z = 1$ for $K = 0.0503$. Two complex branches leave the unit circle for $K = 1.82$ at $z = 1.0\epsilon^{\pm j1.908}$.

PC.9 Two complex branches leave the unit circle at $z = 1.0\epsilon^{\pm j0.567}$ for $K = 0.925$. The other branch remains inside the unit circle for all $K > 0$.

D *MATLAB Commands Used in This Book*

PREVIEW

This appendix contains a summary of most MATLAB commands, including those of the Control System Toolbox and the RPI functions, that are used in the MATLAB scripts throughout the book. This summary does not include the more commonly used commands, such as `plot`. To obtain more information on a specific command, enter `help` followed by the command name, e.g., `help bode`. Other sources for help with MATLAB commands are the manuals, the documentation CD, and the MathWorks web site.

MATLAB COMMANDS

Command	*Purpose*	*Toolbox*
*	Given two LTI objects, the * operator forms their series connection.	Control System
+	Given two LTI objects, the + operator forms their parallel connection.	Control System
abs	Given a complex number, vector, or array, **abs** returns the magnitude(s) of the argument.	MATLAB
angle	Given a complex number, vector, or array, **angle** returns the phase angle, in radians.	MATLAB
axis	The **axis** function allows the user to specify the plotting area. The **axis equal** function forces uniform scaling for the real and imaginary axes.	MATLAB
bode	Given an LTI model, **bode** returns the magnitude and phase of the frequency response. When the output variables are omitted, **bode** generates the Bode plot directly.	Control System
bodedb	Given a single-input, single-output LTI model, **bodedb** returns the magnitude in dB and the phase of the frequency response. The output variables are regular arrays.	RPI function
bwcalc	Given the magnitude of the frequency response of a system and its low-frequency gain, **bwcalc** computes the bandwidth.	RPI function
c2d	Given a continuous-time system and a sampling period T_s, **c2d** converts the system into a discrete-time system.	Control System
canon	Given a system in state-space form, **canon** returns its modal form, with the **'modal'** option.	Control System
clabel	The **clabel** command adds height labels to contour plots.	MATLAB
contour	The **contour** command is used to draw 3-dimensional contour plots.	MATLAB
conv	Given two row vectors containing the coefficients of two polynomials, **conv** returns a row vector containing the coefficients of the product of the two polynomials.	MATLAB
cos	Given a scalar, vector, or matrix, **cos** returns its cosine.	MATLAB
cpole2k	Given a complex pair of transfer-function poles in the z-plane, the residues at those poles, and a discrete-time vector, **cpole2k** returns the time response due to those poles.	RPI function
ctrb	Given the **A** and **B** matrices of a state-space model, **ctrb** returns the controllability matrix.	Control System

Command	*Purpose*	*Toolbox*
damp	Given a model of a discrete-time system as an LTI object, **damp** calculates the equivalent s-plane natural frequencies and the damping ratios of the system poles. When invoked without output variables, a table of poles, damping ratios, and natural frequencies is displayed.	Control System
dcgain	Given a model as an LTI object, **dcgain** returns the steady-state gain of the system.	Control System
deconv	Given two polynomials as row vectors, **deconv** returns the quotient and remainder of the first polynomial divided by the second.	MATLAB
dplot	The **dplot** function plots discrete-time signals in a manner similar to the stem command, but uses triangles as the marker symbol.	RPI function
dzline	The **dzline** function draws lines of constant damping ratios for a discrete-time system.	RPI function
eig	Given a square matrix, **eig** computes its eigenvalues and eigenvectors.	MATLAB
exp	Given a complex number, vector, or array, **exp** computes the exponential function of the elements.	MATLAB
feedback	Given the models of two systems as LTI objects, **feedback** returns the model of the closed-loop system, where negative feedback is assumed. An optional third argument can be used to handle the positive-feedback case.	Control System
fft	Given a discrete-time signal as a vector, **fft** computes the discrete Fourier transform (DFT) of the signal.	MATLAB
find	The **find** command returns the indices and values of the nonzero elements of its argument, which may be a logical expression.	MATLAB
findobj	Given a set of handle-graphics objects, **findobj** returns the handles of those objects having the specified property.	MATLAB
get	The **get** command returns the properties of an LTI object.	MATLAB
hold	When set "on," **hold** draws subsequent plots on the current set of axes. The **hold off** command reverses this action.	MATLAB
ifft	Given the spectrum of a discrete-time signal as a vector, the inverse discrete Fourier transform **ifft** computes the time-domain signal.	MATLAB
imag	The **imag** function returns the imaginary part of its argument.	MATLAB
imp_inv	Given a continuous-time system and a sampling period T_s, **imp_inv** converts the system into a discrete-time system using the impulse invariant transformation.	RPI function

Command	*Purpose*	*Toolbox*
impulse	Given an LTI object of a discrete-time system, **impulse** returns the response to a unit-impulse input.	Control System
interp1	Given two vectors x and y that define the function $y(x)$ and a value x_1, **interp1** returns the interpolated value $y_1 = y(x_1)$.	MATLAB
inv	Given a square matrix, **inv** returns its inverse.	MATLAB
kstats	Given a discrete-time step response, **kstats** finds the percent overshoot, peak time, rise time, settling time, and steady-state error.	RPI function
length	Given a vector, **length** returns its length.	MATLAB
log	Given a scalar, vector, or matrix, **log** returns its base 10 logarithm.	MATLAB
log10	Given a scalar, vector, or matrix, **log10** returns its natural logarithm.	MATLAB
logspace	The **logspace** function generates vectors whose elements are logarithmically spaced.	MATLAB
lsim	Given a discrete-time system as an LTI object, a vector of input values, a vector of discrete time points, and possibly a set of initial conditions, **lsim** returns the time response.	Control System
ltimodels	The **ltimodels** function gives general information about the various types of LTI models supported in the Control System Toolbox.	Control System
margin	Given a model as an LTI object, **margin** returns the gain and phase margins and the crossover frequencies. When the output variables are omitted, **margin** generates a Bode plot with the margins and crossover frequencies indicated on the plot.	Control System
max	Given a vector, **max** returns the largest element. For a matrix, the function returns a row vector containing the maximum element from each column.	MATLAB
mesh	The **mesh** command produces colored 3-dimensional plots, where the color is proportional to mesh height.	MATLAB
minreal	Given a system expressed as an LTI object, **minreal** returns a minimal realization of that system, with coincidental poles and zeros cancelled.	Control System
norm	Given a vector or a matrix, **norm** computes its norm.	MATLAB
nyquist	Given a model as an LTI object, **nyquist** returns the real and imaginary parts of the frequency response. When the output variables are omitted, **nyquist** generates the Nyquist diagram directly.	Control System
obsv	Given the **A** and **C** matrices of a state-space model, **obsv** returns the observability matrix.	Control System
ones	The **ones** command returns a vector or matrix all of whose elements are unity.	MATLAB

Command	*Purpose*	*Toolbox*
pi	This is the constant $\pi = 3.14159\ldots$	MATLAB
place	Given the **A** and **B** matrices of a state-space model, **place** returns the gain matrix **F** that places the eigenvalues of $\mathbf{A} - \mathbf{BF}$ at specified locations in the z-plane.	Control System
pole	Given an LTI object, **pole** computes the poles of the system's transfer function.	Control System
pzmap	Given an LTI system model, **pzmap** produces a plot of the system's poles and zeros in the z-plane.	Control System
rank	Given a matrix, **rank** determines the number of linearly independent rows or columns in the matrix.	MATLAB
real	The **real** function returns the real part of its argument.	MATLAB
reg	Given state-feedback and estimator gains, **reg** produces an observer-based regulator as an LTI object.	Control System
residue	Given a rational function $T(z) = N(z)/D(z)$, **residue** returns the roots of $D(z) = 0$, the partial-fraction coefficients, and any polynomial term that remains.	MATLAB
rlocfind	Given a TF or state-space description of an open-loop system, **rlocfind** allows the user to select any point on the locus with the mouse and returns the value of the loop gain that makes that point a closed-loop pole. It also returns the values of all the closed-loop poles for that gain value.	Control System
rlocus	Given an LTI model of an open-loop system, **rlocus** produces a root-locus plot that shows the locations of the closed-loop poles in the z-plane as the loop gain varies from 0 to infinity.	Control System
roots	Given a row vector containing the coefficients of a polynomial $P(z)$, **roots** returns the solutions of $P(z) = 0$.	MATLAB
rpole2k	Given a real transfer-function pole in the z-plane, the residue at that pole, and a discrete-time vector, **rpole2k** returns the time response due to that pole.	RPI function
semilogx	The `semilogx` function generates semi-logarithmic plots, using a base-10 logarithmic scale for the x-axis and a linear scale for the y-axis.	MATLAB
set	Given a handle-graphics object, **set** assigns the value of the specified property.	MATLAB
sin	Given a scalar, vector, or matrix, **sin** returns its sine.	MATLAB
small20	The **small20** function replaces very small matrix elements by `0`.	RPI function
sqrt	The **sqrt** function returns the square root of its argument.	MATLAB
squeeze	Given an array with one or more singleton dimensions, **squeeze** returns an array with the same elements, but with all the singleton dimensions removed.	MATLAB

Command	*Purpose*	*Toolbox*
ss	Given a set of state-space matrices or an LTI model in TF or ZPK form, **ss** creates the model as an SS object.	Control System
ssdata	Given an LTI model, **ssdata** extracts the state-space matrices **A**, **B**, **C**, and **D**.	Control System
stairs	The **stairs** function draws a stairstep graph of the vector's elements.	MATLAB
stem	Plots the data sequence Y as stems from the x-axis terminated with circles for the data value.	MATLAB
step	Given an LTI object of a discrete-time system, **step** returns the response to a unit-step function input.	Control System
subplot	The **subplot** command allows the plotting window to be divided into multiple plotting areas.	MATLAB
tan	The **tan** function returns the tangent of its argument.	MATLAB
text	Adds text to plots at the specified location.	MATLAB
tf	Given numerator and denominator polynomials or an LTI model in SS or ZPK form, **tf** creates the system model as a TF object.	Control System
tfdata	Given an LTI model, **tfdata** extracts the numerator and denominator polynomials.	Control System
tzero	Given an LTI object, **tzero** returns the zeros of its transfer function.	Control System
ucircle	The **ucircle** function plots the unit circle.	RPI function
vgain	Given a model as an LTI object, **vgain** computes the velocity-error constant.	RPI function
xy2p	Given a complex variable, **xy2p** returns its magnitude and phase (in radians).	RPI function
zeros	The **zero** command returns a vector or matrix, all of whose elements are zero.	MATLAB
zgrid	When viewing either a root-locus plot or a pole-zero map in the z-plane, **zgrid** draws contours of constant damping ratio (ζ) and natural frequency (ω_n) within the unit z-plane circle.	Control System
zpk	Given a system's zeros, poles, and gain, or an LTI model in TF or SS form, **zpk** creates the system model as a ZPK object.	Control System
zpkdata	Given an LTI model, **zpkdata** extracts the zeros, poles, and gain.	Control System

ANNOTATED BIBLIOGRAPHY

1. Apkarian, J. *A Comprehensive and Modular Laboratory for Control Systems Design and Implementation.* Toronto, ON: Quanser Consulting Inc., 1997.

2. Franklin, G. F., J. D. Powell, and M. L. Workman. *Digital Control of Dynamic Systems*, 3rd ed. Reading, MA: Addison-Wesley, 1997.

3. Franklin, G. F., J. D. Powell, and A. Emami-Naeini. *Feedback Control of Dynamic Systems*, 3rd ed. Reading, MA: Addison-Wesley, 1994.

4. Frederick, D. K. and J. H. Chow. *Feedback Control Problems Using MATLAB and the Control System Toolbox.* Pacific Grove, CA: Brooks/Cole, 2000.

5. Hanselman, D. C. and B. R. Littlefield. *Mastering MATLAB 6: A Comprehensive Tutorial and Reference.* Englewood Cliffs, NJ: Prentice-Hall, 2001.

6. Houpis, C. H. and G. B. Lamont. *Digital Control Systems, Theory, Hardware, and Software*, 2nd ed. New York, NY: McGraw-Hill, 1992.

7. Jacquot, R. G. *Modern Digital Control Systems*, 2nd ed. New York, NY: Marcel Dekker, 1995.

8. Kuo, B. C. *Digital Control Systems*, 2nd ed. Fort Worth, TX: Saunders, 1992.

9. The MathWorks. *Using MATLAB, Version 6.* Natick, MA: The MathWorks Inc., 2000.

10. The MathWorks. *Control System Toolbox User's Guide, Version 5.* Natick, MA: The MathWorks Inc., 2001.

11. Ogata, K. *Discrete-Time Control Systems*, 2nd ed. Englewood Cliffs, NJ: Prentice-Hall, 1995.

12. Oppenheim, A. V. and R. W. Schafer. *Discrete-Time Signal Processing,* 2nd ed. Englewood Cliffs, NJ: Prentice-Hall, 1999.

13. Phillips, C. L. and H. T. Nagle. *Digital Control System Analysis and Design*, 3rd ed. Englewood Cliffs, NJ: Prentice-Hall, 1995.

Index

J

K

L

M

N

O

P

R

S

T

U

V

W

Z